White Feathers

WHITE FEATHERS

The Nesting Lives of
Tree Swallows

BERND HEINRICH

Houghton Mifflin Harcourt
Boston   New York
2020

For information about permission to reproduce selections from this book, write
to trade.permissions@hmhco.com or to Permissions, Houghton Mifflin Harcourt
Publishing Company, 3 Park Avenue, 19th Floor, New York, New York 10016.

hmhbooks.com

Library of Congress Cataloging-in-Publication Data
Names: Heinrich, Bernd, 1940– author, illustrator.
Title: White feathers : the nesting lives of tree swallows / Bernd Heinrich.
Description: Boston : Houghton Mifflin Harcourt, 2020. |
Includes bibliographical references and index.
Identifiers: LCCN 2019024927 (print) | LCCN 2019024928 (ebook) |
ISBN 9781328604415 (hardcover) | ISBN 9781328603517 (ebook) |
ISBN 9780358172338 | ISBN 9780358306887
Subjects: LCSH: Tree swallow. | Tree swallow — Nests.
Classification: LCC QL696.P247 H45 2020 (print) | LCC QL696.P247 (ebook) |
DDC 598.8/26 — dc23
LC record available at https://lccn.loc.gov/2019024927
LC ebook record available at https://lccn.loc.gov/2019024928

Printed in the United States of America
DOC 10 9 8 7 6 5 4 3 2 1

CONTENTS

Acknowledgments ix

Introduction xi

2010
1. STARVED IN THE NEST 1

2011
2. DEVOTION TO THE LAST FLEDGLING 9

2012
3. A SUCCESSFUL CLUTCH 21

2013
4. CLUTCH SIZE AND WEATHER WIPEOUTS 25

2014
5. NESTING NOVICE 49
6. EXPERIMENTS WITH FEATHERS 63
7. TENDING THE BABIES 75

2015

8. A NESTLING REDUCTION 89

2016

9. SERIAL TENANCY AT THE NEST-BOX 105
10. MATING TIME 113
11. EGGS AND MORE FEATHERS 131
12. DISAPPEARING CHICKS 139

2017

13. DOMINATING OF THE AIRSPACE 147
14. FEATHER PLAY 157
15. EGG LAYING AND INCUBATION 163
16. LAST BUT NOT LEAST 169

2018

17. THE APPLE TREE PAIR 187
18. LATECOMER'S ADVANTAGE 201

Postscripts 209

The Literature 219

Index 227

Not to be confined by the greatest, yet to be contained within the smallest, is divine.

— Epitaph of St. Ignatius (1491–1556)

I was walking across our compound last month when a queen termite began building her miraculous city. I saw it because I looked down. One night three great fruit bats flew over the face of the moon. I saw it because I looked up.

— William Beebe (1877–1962)

ACKNOWLEDGMENTS

I thank Ted Simanek for building my nest-boxes to swallow-friendly specifications, and I thank him and Betty Simanek, as well as Linda Bean and Nancy and Richard Stowell, for allowing me to distribute them over their fields. Willem Hillier too for aerial pictures of the study area with a drone. Nathanial T. Wheelwright, Margaret McVey, Liam Taylor, Sandra Mitchell, and Paul R. Spitzer shared observations of other swallows, and Margaret McVey made helpful comments and suggestions on the manuscript. Special thanks to Lynn Jennings for her patience and understanding of my long and frequent absences, not only during my observations of the swallows in the field, but during continuing entanglement afterward. I might not have done it with as much abandon but for the thought that she would always still be there, and I am sorry for unanticipated costs incurred. As always, I greatly appreciate the staff at Houghton Mifflin Harcourt and their faith in me, and this project. I especially thank Deanne Urmy and Susanna Brougham for their careful reading, excellent suggestions, and patience.

INTRODUCTION

THERE IS ARGUABLY NO BIRD IN THE WORLD THAT COMbines graceful flight, beauty of feathers, pleasing song, and accessibility, plus tameness and abundance, more than the tree swallows (*Tachycineta bicolor*). And just by putting up a nestbox made in minutes from some scrap board and placed on a pole, I had a pair nesting by my door. In early May 2008, I happened to peek into the nest-box and saw five snow-white eggs in a bed of long white feathers. I had peeked into nest-boxes before and seen nest linings of various commonly available materials, but never anything like this. It was no fluke — such white feathers are rare, and it had cost the swallows deliberate effort to search for and acquire them.

Not surprisingly, swallows are among the most studied birds, and the tree swallow has been considered a "model" bird for research, as the fruit fly (*Drosophila melanogaster*) and the house mouse (*Mus musculus*) are to genetics, and the Norway rat (*Rattus norvegicus*) is to animal behavior. In the many hundreds of scientific studies of tree swallows, the most common topic is

their mating behavior. But when I wanted to find out why they line their nests with white feathers, there was nothing in the literature. I had little interest in the mating behavior of tree swallows, but why they went to the effort of finding white feathers was an intriguing question. Would this behavior be repeated by the pair near my door, or by other pairs, or at another place? No answers to these questions were readily available, and I was sure that many more unknowns were linked to them.

The species is distributed all over North America. Originally tree swallows nested only in tree holes, in contrast to the perhaps more familiar barn and cliff swallows, which build their own type of birdhouse by plastering one gobbet of mud at a time onto a solid substrate to make a potlike cavity, and there they nest. Theirs is a wonderfully creative and beautiful innovation, which allows them to pitch their house potentially anywhere with a convenient solid surface, such as a barn beam or a cliff face. They can nest directly next to each other, and sometimes even on top of each other, thus sharing walls and saving construction costs. As many as fifty cliff-swallow nests may be found in a square meter. In contrast, tree swallows are normally solitary during nesting time. They nest in holes made by woodpeckers, and since woodpeckers are territorial, commanding a certain area as their own, these nesting sites are widely distributed.

But we can bring tree swallows closer to us. By providing them with substitutes — ready-made nest-boxes — we'll find them nesting right in front of us, in plain view. Their bubbly songs and lively activities are reminders of life's beauty and bounty, and sometimes their presence provides the opportunity for close study.

Henry David Thoreau famously claimed that a closely examined life would yield infinite riches. He meant our individual lives. But why not consider others'? Birds live by needs, means, and constraints similar to our own. Like us, they claim and settle in a neighborhood, secure food, court a mate, build a home, raise their young, and avoid the dangers that could imperil their own lives and those of their offspring. The details of how they accomplish those universal tasks in their world might offer perspective on and insight into our own lives. Ordinarily we barely glance at swallows; I wanted to watch them deliberately and get to know them intimately.

As you will see, I have attempted, with as little interference as possible, to follow the life cycle of one pair of tree swallows per nesting season. Occasionally, prompted by my observations,

I explore wider topics. This differs from the much-practiced method of studying large numbers of birds at a time. My focus is on the detailed observation of individuals, and I have included both drawings and photographs so readers may take this journey with me through ten nesting seasons marked by repetition and variation. My hope is to present an intimate view of the nesting life of a fascinating species.

White Feathers

STARVED IN THE NEST — 2010

IN 2010, WHILE LIVING IN VERMONT, I OFTEN EXAMINED nest-boxes being used by various pairs of birds — house wrens, great crested flycatchers, European starlings, black-capped chickadees, and tree swallows. The swallows had won out over a pair of chickadees in a contest for the same box, and they furnished the nest with white feathers but, curiously, did so only *after* their eggs had been laid — not before, as per usual avian protocol.

The female began incubating her six eggs on May 28, 2010, and the nest eventually contained 110 white feathers. The eggs hatched on June 8, and by the 24th the young were feathered out in their ash-gray garb. At that point the nestlings were clambering up from the bottom of the nest to perch, one at a time, in the nest-box entrance hole. There they intercepted each parent as it brought food, fluttering up to its offspring, sometimes hardly stopping. The adult bird simply passed off the food while on the wing.

As this was going on, the adult swallows ignored me and Hugo, my yellow Labrador mutt. But on June 26, they suddenly became noisy and started to dive at Hugo and me, and even made a pass at a kingbird that was pulling threads out of a now-empty oriole nest in the neighboring tall black-cherry tree. When a blue jay landed in the cherry tree near the nest-box, one of the swallows dived at it while noisily chattering an alarm.

Then, for the first time, I saw the male take a short break from his usual nest-tending routine; he landed (twice) on the nest-box, his head and back feathers glistening greenish blue to black, depending on the angle of light. I had never seen one of the swallows stop to land there since after the eggs had been laid, but now both of them landed, and also hovered in front of the nest entrance, but without passing off food. This new behavior was puzzling because earlier, the two had consistently come singly and flown directly to and from the box entrance, always to deliver a meal.

Early the next day, the 27th, the baby leaning out of the nest-box was cheeping nonstop, and at 10:10 a.m. it leaned out and almost fell, but, fluttering hard, managed to hold on to the edge of the box by its feet. It tried to get back in, but another nestling had by then taken its place and was blocking the entrance, so, unable to hang on, the first baby fell straight down into the grass and weeds, a distance of two meters. It sat still there for a few minutes before clambering, with rapid flutters, back up into the vegetation. It then perched on the garden fence directly underneath the nest-box.

A Cooper's hawk came out of the woods and perched in a tree nearby. It had, I suspected, been attracted by the baby swallows' hunger cries — the continuous two-syllable *chil-it* calls, which had been increasing in volume. The baby swallow was a "sitting duck" that would, I thought, have been taken by now, had I not been near enough to discourage the hawk.

The pair of swallows then resumed flying to and from the box, but I had the impression they were feeding the babies less frequently. Could that really be true? To make sure, I needed numbers, and soon found they were making one trip, on average, every 3.5 minutes, which seemed about half as frequent as before. The baby that had fallen out of the nest continued meanwhile to

call out like the others. It also preened, shook itself, and then to my surprise flew off in seemingly competent flight, landing in a nearby ash tree.

The Cooper's hawk returned later, making a quick flyby, and a broad-winged hawk soared overhead. One of the adult swallows chittered loudly and flew up to meet it, looking tiny next to the high-flying hawk. But the swallow was safe, being more agile in flight than any hawk.

The volume of the young swallows' peeping increased, and now another leaned farther out of the nest-box to meet its approaching parent, even when that bird was still a hundred meters away. However, the young swallow did not react to other birds flying near. Apparently it could recognize its parent. This too was surprising to me.

The begging of the swallow chicks grew more strident as the food deliveries decreased in frequency, but the babies were unable to stimulate their parents to return faster. Feeding nearly came to a stop from 2:30 to 5:48 p.m.; only one food delivery took place during that period of more than three hours. Except during bad weather, I had never before observed such a long interval without food coming to the nest. Two minutes after the 5:48 p.m. feeding, one parent came back, but not to feed the young. It flew round and round, and back and forth, by the nest-box, and called for seven minutes. On previous days neither parent had hesitated a moment before going to the nest entrance. If this change in behavior was meant to lure the babies out to fledge, it wasn't working so far; by 8 p.m. all but one of the babies still remained at the nest, and they now reacted to the parents when they came within a range of two hundred meters.

June 28 started warm (68ºF), wet, and windy. It had poured

during the night but the rain had stopped by the time I got up. Convinced that the remaining five young would fledge today, I had made a special effort to get up in the dark to try to find out what role, if any, the parents might have in this event, and whether the family would leave the nest together. At dawn, already long after the robin had belted out its song and after I had let my two young Canada geese out of their pen and they had run after me to settle beside me in the grass, I started my watching. But there was no sound from the tree swallows. A female American robin sat tight on her nest in a birch tree in front of me. A rose-breasted grosbeak and a warbling vireo sang from their usual trees. A male ruby-throated hummingbird perched beside me on a dry raspberry stem, as he often did, and a mourning dove sang his haunting song. Not knowing when the fledging would occur, I expected an all-day wait, and after a while I settled in on a comfortable patch of grass, distracted by cedar waxwings.

The waxwings were now dissolving their small flocks. Five birds flew by, and two peeled off to their nest near the end of a long horizontal pine limb. Two others engaged in a chase against the fifth. A little later one was vibrating its wings and opening its bill in typical food-begging behavior, as the other offered it a bright-red honeysuckle berry. It was almost nesting time for these identically garbed birds, and likely they were following the usual bird protocol: a male was offering a female a so-called nuptial gift. I presumed it was a male suitor. The female accepted the berry, and then begged again. This time he *acted out* feeding her, though he had no berry in his bill. He pretended to pass her one; it looked like a kiss.

The pair of song sparrows that had fledged their young several days earlier had already started a new nest near my watching post. The male sang, and the female flew with one piece of dry grass after another, at intervals of about a minute, into the dense patch of weeds in front of me. He accompanied her, but when she flew down to the nest, he flew up into a tree and sang a few soft, muted refrains — mere whispers compared to those he belted out in April, the day after he arrived.

There was no activity at a nearby nest of northern orioles, which was hanging from a bough of the large overarching black-cherry tree; the young had long since fledged. The strands of milkweed the birds had gathered to make their nest had by now been reused in one kingbird and two cedar-waxwing nests. A family group of four ravens drifted by, circling lazily in the air, dipping down, flying back up. Their young continually yelled — pitiably, I thought — as they tagged along. I had never let young ones (ravens or others) in my care get quite so hungry that they screamed so. Families of blue jays and common grackles were also near, and the young of each were making a constant racket. The grosbeak sang, with only small breaks, and the female low in the bushes below his perches fed on both honeysuckle and serviceberries. Yesterday the male had followed her closely as she wandered through the underbrush; he was either silent or singing in a barely audible whisper. That is, this song was meant for her specifically — it was not a full-throated proclamation that this was the bird's territory. It was intimate music for her ears only, now that the two were a pair close to nesting.

Meanwhile, this morning, one or another of the baby swallows had perched at the nest-box entrance. It seemed that the parents were still trying to starve them out because so far no feeding had occurred, and even by 7 a.m., when the female came by for the first time, she had no food. She merely landed briefly by the nest entrance and then left.

The swallow parents made two more flybys over the next hour and a half, and the baby at the nest entrance, reaching farther and farther out, peeped piteously. Finally, at 9:51 a.m. one of the parents flew up to it and passed it a morsel of food, and for the next nine minutes this parent flew around the nest area without a break, fluttering, gliding, diving, making sharp turns, starting to fly off but coming back again and again, calling all the while. Two more visits to the nest within the next hour were followed by similar display flights. By 11 a.m. there had been only seven visits from the parents.

A second young swallow flopped over and hung on at the nest entrance, fluttering, as one had done yesterday. However, after managing to retract its wing, it squeezed back into the nest-box.

Finally, at 6:42 p.m., another baby swallow flew away. It left alone, with no adults near at the moment. The young bird fluttered clumsily along and kept losing altitude, heading downward toward thick raspberry bushes. But it then pedaled hard and barely cleared them, and continued toward the nearby beaver bog. I feared it would land in the water, but it turned and, "finding its wings" in time, circled and headed for a pine tree. At that point in its exodus an adult swallow suddenly arrived and accompanied it. Baby and adult then flew together out of my sight. Greatly relieved, I was ready to examine the contents of the now-silent nest-box. I opened it.

The white feathers were now trodden into a damp, gooey mass. Flat on the layer of guano lay a fully feathered, fully dead young swallow. I could see no injury, but the keel of its chest was sharp; it had starved and was now in rigor mortis, so it had died within the past day. Other than that, the nest was empty.

None of the young returned there. I could not be sure how many had survived; I'd seen two leave the nest but could have missed others as they did so. But for most of the next two days, tree swallows were skimming over the pond and over neighboring fields, and my curiosity was whetted. What were these birds capable of, and how did they solve the basic problems of living?

Six is usually the maximum number for a clutch of tree swallows. Perhaps this time the pair had laid one extra egg, or something else had happened. Another female may have discovered the nest and inserted her own egg there at an opportune moment.

2

DEVOTION TO THE LAST
FLEDGLING — 2011

THE TREE SWALLOW NEST I HAD OBSERVED IN VERMONT showed possible advantages as well as difficulties related to using tree holes as nesting places. The nest holes were small — perhaps all the babies hadn't been fed because not all could perch together at the entrance. Now, with my move in 2011 to our remote cabin in the Maine woods at the center of a half-hectare clearing, a place where I had spent many previous summers, I could definitely rule out any white feathers coming from a nearby chicken farm. Here in Maine, there were no farmers for miles. My first surprise, though, was to see swallows battling.

SWALLOWS ARE KNOWN for their peaceful, mild-mannered nature. A barn with barn swallows is always alive with their gentle twittering, as they sail in and out the door or an open window and perch on their nests plastered here and there on the ceiling beams. I don't recall ever seeing a clash or quarrel among barn, cliff, or bank swallows. Not so here in our clearing, with nine nest-boxes ready for occupancy.

Two swallows had come as a pair, and the first thing they did upon arrival was to land repeatedly at the entrance of the same nest-box. They flew to and from it, stopping to peek in, then flying around the clearing while chirping loudly, only to return again and again, looking in some more. They ignored the other newly installed boxes, suggesting that one or both of these swallows were returning to an old nest site. Then, within an hour, a third swallow arrived.

Within less than a minute, one of the other two swallows met the stranger in the air. The two then fell to the ground, entangled in a brawl that looked like a single ball of fluttering feathers. They were close by, oblivious to my presence, so I walked over. I was almost able to pick up the grappling duo before they disentangled themselves and flew up into the air. Was this an extraordinary event, or was it characteristic of tree swallows?

Only the original pair remained after several such battles with the third. But fifteen minutes later, suddenly four swallows were flying around the clearing. The new pair left after being chased off, but this time no contact fights occurred. One of the two remaining swallows then perched on top of a nest-box, while the other clung to its entrance and peeked in.

Two weeks of cold and rainy weather followed this encounter, and during this time I saw no swallows. However, on May 7, temperatures rose to 63 degrees F, and masses of migrant birds came streaming in. Already at dawn, four swallows were circling the clearing, but only one pair remained — apparently that first pair from two weeks earlier. The female (identifiable by her dull brownish coloration, distinct from the shiny reflective blue of the male) carried a bit of grass into the nest-box the two had apparently claimed on their first day here.

A week later the nest of grass looked finished, and egg laying started the next day, May 15, even though the nest still had no lining of feathers. By May 19, with egg laying nearly finished, it contained six light-colored feathers. Examining the feathers closely, I identified them as mostly those of ducks. The nearest potential sources were the beaver bog two kilometers to the south and the lake twice as far to the west.

I knocked on the nest-box in early June when the eggs were being incubated, hoping the bird inside would flush out so I could examine the eggs. But there was no response, and so I opened the nest-box by removing its front panel. The female was on the nest. She didn't budge. She seemed nonchalant as I gently reached in and poked a finger under her belly to shove her gently to the side as I counted six eggs. After that I checked the nest daily and, as before, she appeared unfazed by my presence.

She and her mate continued to ignore me. I had probably in an earlier summer become an accepted part of their environment. Swallows, like people and most other animals, are alarmed by the unfamiliar and the unexpected. I was not news to these swallows. With my cabin door practically next to their nest-box, I was the equivalent of the rain barrel under the roof or the wheelbarrow by the garden.

The six eggs hatched on June 10. Young songbirds typically put on conspicuous growth from one day to the next, which allows them to become quickly independent — sometimes merely ten days after hatching. By June 27 all six young were feathered out, and they begged nonstop. The baby perched in the nest entrance increased the volume of its calls whenever one of the adults flew into the clearing.

I expected the young to fledge soon, so the next day, June 28, I closely followed the adults' nest-provisioning behavior. They now averaged only five nest visits within ten minutes, one-third the frequency that I had measured earlier. The reduction could have been due to any of a number of variables, such as weather, hatches of specific insects, and so on. But the adult birds' activity soon seemed deliberate, by choice: in the last two minutes of one of my ten-minute watches, one of the birds circled around the whole time. It cheeped emphatically while swooping repeatedly over and around the nest-box, but without landing. It had an agenda other than food delivery.

The next morning the familiar constant begging calls of the baby swallows continued unabated. I knew that six fledglings had so far survived and were still in the nest. But curiously, no baby swallow was now perched in the nest entrance; up to this moment, one or another had been there virtually continuously. Furthermore, instead of two adults, only one now circled the clearing, making soft musical calls. It kept on and on with this, until finally, alighting at the nest entrance, it perched there. By the cadence of the sound, I could tell that only one baby bird was now begging, and even more incessantly; but it still did not appear at the nest entrance. I had, by taking a short break in my watch, missed the apparent quick departure of all except one baby.

I again monitored the parents' frequency of visitation, and during ten minutes saw thirteen visits to the nest-box entrance. But despite this sudden increased attentiveness, no food was passed to the young — despite the fact that during at least the last six to eight visits, a parent bird carried a conspicuously large food item: a yellow-brown moth. I had seldom seen the swallows carrying any sort of large insect, and I had never before observed them bringing in prey that large. I had the impression that the swallow was using it as bait in an attempt to lure the last baby out of the nest-box. But something wasn't right — only one parent was present at any given time, whereas usually I had seen both together. Furthermore, no baby had appeared at the entrance, despite the constant begging. I therefore approached the nest-box to check its contents. The begging stopped, and when I opened and looked inside, I was surprised. The nest was empty! I looked again and again, at first not believing my eyes.

No baby could hide in that nest. It was a soggy, stinking, thick layer of guano. I looked under the guano — it was crawling with the usual guano-feeding maggots, generally harmless to birds. The conclusion was clear: although the babies had fledged, one remained nearby.

The moment I left the vicinity of the empty nest-box, the begging resumed, coming, I now realized, from a thicket of blooming fireweed directly beneath the box. I then found the baby on the ground there, under a meter-thick layer of dense plants. No adult swallow would enter such a tangle, nor would any juvenile, unless it had fallen into it. I picked up the baby swallow and returned it to the box, hoping the parents might still feed it.

One parent returned twice to the nest-box during the next half-hour. But each time the parent (always just one) circled and called; it did *not* feed the one remaining nestling, which now perched in the entrance. Seeing the adult flying nearby, the baby responded by calling ever more wildly, until it finally jumped out. It was still too weak to fly, and again it fell into the fireweed.

In ten minutes a swallow landed at the nest-box entrance, and the baby below screamed as loudly as ever. But the parent soon left. Forty-seven minutes later, a parent once more returned. This time it did not fly to the nest-box but instead made several brief passes high over the clearing before flying on. There was no response from any of the other five babies anywhere within earshot. They had apparently all left with one of the parents; one of the pair kept returning to try to retrieve the one remaining baby, which was weak. Now fearing that this nestling might be abandoned, I retrieved it and saw that it was fully feathered out, but its pectoral (flight) muscles were atrophied — the bird was starved. I tried to revive it.

Nearby flowers offered many insects. Prying open the baby's bill, I was able to insert crush-killed insects, mostly bees — they were the most abundant and easy to catch. The bird eagerly ingested all that I gave it. (I did not want it to get stung by the honeybees — even recently dead honeybees still have a functioning stinger. I held them to my arm first to discharge the venom, which simultaneously stimulated my immune system.)

While I was walking around, capturing and preparing the food for my new charge, two swallows, presumably *both* parents, suddenly reappeared. The youngster instantly started to struggle in my hand, so I released it. Now with a clear view ahead of

it, and with a little fuel in its stomach, it fluttered downhill until it reached the woods at the bottom of the slope, about a hundred meters away, where it landed in a low bush. It there resumed calling. The adults instantly flew to it. They did not, however, feed it. They had come only to look for it, and just swooped over it and left.

A half-hour later *three* adult swallows arrived, but just one stayed, for twelve minutes. Seventeen minutes later two came. Still, none fed the baby. I again retrieved the little bird; its calls made it easy to locate. It made no attempt to fly as I plucked it from its perch and put it into a makeshift wire cage, which I set on the ground near the nest site.

The pair came by several more times that still-early afternoon, and they now landed at the cage. I had by then fed the youngster ten insects and felt it must be absolutely stuffed. However, since the parents' presence still invariably caused it to cheep loudly, I wanted to find out whether it was hunger or their presence as such that caused it to respond like that. I therefore fed it until it would take no more, to find out if it would still cheep at the parents when they came by. The baby eventually ingested, one after another, at least thirty-eight large insects (three beetles, four honeybees, twenty-one flies the size of honeybees, eight bumblebees, and two butterflies). Afterward it sat at rest and seemed content and silent for an hour. Unfortunately, my intended experiment yielded no result; no other swallow came by in all that time.

That evening I set the still-inept fledgling on a branch, where within minutes it was again chirping nonstop. Then, in a weak and clumsy but nevertheless fluttering flight, it flew a hundred meters to a nearby chestnut tree. It gradually got stronger, and from there it flew to a nearby spruce, and then onto a sugar maple.

An adult swallow finally came at dusk, clearly announcing itself by chirping continuously as it circled all around. The baby immediately answered but did not budge from its perch. Eventually the circling swallow switched to higher-pitched calls. The baby continued to respond, but it did not jump off to follow the circling swallow, which after three minutes of continuous circling flew off in silence in a southerly direction, as both adults always had when leaving the clearing.

The baby had now possibly been abandoned. Yet once again a swallow came before dark. This one didn't circle. Instead it flew low and went directly to the baby; then, after a brief vocal exchange and possibly a feeding, the adult left. I had by now noticed two flight patterns, or individual flight signatures — those of the female and male, respectively. One hurried and fluttered, the other with some gliding.

There was silence the next dawn. But at 5:20 a.m., on this clear and windless day, the male came for a reconnaissance. Its calls of *cheelp — cheelp-cheelp* sounded insistent as he flew around the maple tree where the baby had perched the previous evening. The swallow then flew low and called all over the clearing, but this time no cheeps answered him, and he left after a mere fifteen seconds. He had been back long enough, and his calls had been conspicuous enough to elicit a response, if the baby was still there.

Three minutes later the female (the one with the more fluttering flight and the higher-pitched but softer voice) came by also.

She made only two or three passes before also flying off, silent, over the forest in the southerly direction, as her mate had done, and as both had done the day before.

That morning I continued to lie in bed, next to the open window, after the two swallows had left, listening to the dawn chorus of ovenbirds, of yellow-throated and chestnut-sided warblers. A family of common ravens, which had noisily settled at dusk into a nearby pine grove, were just waking up and making a racket loud enough to raise the dead. A loon called from the lake, and an evening grosbeak's clear, bell-like tone rang out, but there was no peep from the baby swallow. It could not have left after dark, I had seen it just before dark, and no baby now responded to either parent. I surmised it had died during the night.

I DID NOT EXPECT to see the parents again this year because they had come to check on their fledgling and heard no response. In past years, after the young had all fledged, no adult swallows appeared, even on the day that the fledging took place. Not until the following spring would another swallow grace the clearing. Yet on *this* morning, the pair showed up five more times, vocalizing to call their young! This, given my previous observations, was noteworthy. It suggested that the parents had inferred that this one baby, the last of six, was still around. I doubt they actually count their young, so the swallows' inference was likely based on having seen the fledgling the previous evening. Yet I was still baffled, because the five other babies had most likely left with the parents — wouldn't the adults have felt they had them all?

After each foray, the pair flew off in a southerly direction, to the place where surely they had taken their other five young,

possibly the beaver bog. The five babies must have left close together, perhaps as a group. I now did know that the parents remembered and cared about the one that got left behind, or else
they would not have come here repeatedly, looking for it. I renewed my efforts to find out what might have happened to the
little swallow. I had seen it at dusk, but there was no sound of
it at dawn; it was therefore not likely taken by a hawk, and so I
searched in the grass and leaves under the tree where the baby
had perched at night. I found no sign of it.

That afternoon, after the swallows stopped coming, a turkey
vulture kept circling. Could a vulture have been attracted by the
scent of a single dead swallow? This new thought induced me
to recheck the thickly overgrown ground under the maple tree,
and this time I found the baby lying dead among the leaves. Its
keel was still sharp, due to greatly atrophied flight muscles, and
its intestines and stomach were empty except for the black mush
of insect exoskeletons. Like the fledgling at the Vermont nest,
it had apparently died of starvation, although this seemed hard
to believe, given how much I had fed it the previous afternoon.
It may, however, have been overfed and died of shock, after the
long period without food.

This was the last swallow I encountered that year. My observations of the family included details that might have seemed
unbelievable, had I not seen and experienced them. The swallows had indicated that they provide incentives to offspring for
leaving the nest, then keep track of individuals, have expectations, vocalize mutually to communicate their presence or location, and remember where individuals should be, in the midst of
changing circumstances. There seemed a deliberateness about
their behavior that, if observed among humans, would likely

be interpreted as understanding. But this was only one pair of swallows; I needed to follow more of them and gather more details before I could parse the instinctual from the habitual, the uniquely appropriate from the random.

But I could at least draw this conclusion. Six offspring were too many for these parents to raise. And the spring climate limits the number of insects available as food. There is then presumably strong selective pressure for a pair of swallows to limit the size of a given clutch. One way to retain a large number of offspring is to deposit eggs in the nest of another family of swallows.

3

A SUCCESSFUL CLUTCH — 2012

A HOT SUMMER, A BEAUTIFULLY COLORFUL FALL, AND A long, cold, and dark winter led to a slow thaw in March. The first swallows did not show up until May 7, about a month later than usual. As in the previous year, a pair arrived, and the two immediately flew directly to the nest-box (out of an available nine) they had used before. When not in the air or at the box itself, the duo perched side by side on the same branch at the top of the black locust tree where they had habitually perched the year before.

Three days later I awoke at dawn, hearing a swallow calling, and jumped out of bed in time to see one flying high over the clearing. It was violently chasing after another. I stood mesmerized, watching what I now realized was three swallows ranging far and wide in swift pursuit that continued for an hour and ten minutes, until the third one finally left. After this the pair began nesting, and over the next five days they mated frequently. On

May 18, the first egg appeared in the nest, and then the usual one egg per day proceeded until stopping at the sixth, the usual maximum number for a clutch.

I kept track of the nest's contents, and the pair remained unperturbed by me. As before there were no feathers in the nest, despite eggs having been laid. Most birds lay their eggs only after they have put in the nest lining. It seemed possible that the swallows had been unable or too busy to find any feathers earlier. But I now provided some — tossing a few feathers onto the ground. One of the pair immediately swooped past me to get them.

Before incubation started, there was often one and sometimes two other swallows nearby, and again chases ensued. But after incubation started, all chases stopped.

Five young hatched on June 6, and the sixth on the next day. The thick layer of fecal material that I had found in the nest the previous year had made me wonder if the swallows practiced nest hygiene. Now, as I kept close watch, I saw both adults carrying out this material in membrane-bound sacs, easily carried in the bill. They dropped these sacs at least a hundred meters from the nest, although I once noticed one popping out of the nest entrance right after one nestling had sat there and received seven consecutive feedings.

The first of the young fledged at 3:49 p.m. on June 26, after leaning out the nest-box entrance more and more, then finally launching itself into the air. It circled several times around the clearing and then landed on the tiptop of a tall pine tree, where a parent joined it. The little bird was chatty, making continuous two-note calls sounding like *chil-it,* one after the other. Then it flew around with one of the adults, its flight seemingly executed to perfection at the first try. It left the clearing after only twelve minutes out of the nest, accompanied by a parent. Two more of the young left at about 7 a.m. the next day, and they landed in a nearby patch of fireweed, where they continued to vocalize; both parents made trilling-tinkling calls while circling over them.

These three were the only young I happened to see leave; there were gaps of time when I was not watching the nest-box entrance. But when I heard no more babies cheeping from the nest-box, I opened it and found it empty; all six had indeed fledged. The two recently fledged babies perched in the fireweed by the nest-box, preening themselves. At 9:16 a.m. a parent returned, trilling at high pitch and also making tinkling calls. The young answered the adult, and within a minute the last babies of the clutch launched into flight and followed this parent in the well-established southerly direction. This nesting closely repeated the pattern I had observed in the previous two years, except there was a larger margin of safety for the offspring. The last two of the young would likely make it, provided they had enough fat to supply energy until they could feed themselves independently. The later date of this year's nesting may have made the difference.

4

CLUTCH SIZE AND WEATHER WIPEOUTS — 2013

TREE SWALLOWS USUALLY RETURN MORE THAN A MONTH before they begin to nest, when snow may still be in the woods and few flying insects are available as food. But despite the very early spring arrival, they raise only one clutch of young per year. The local barn swallows and woodland birds that return a month later raise two or more clutches of young per year. So why do tree swallows rush back to nest before abundant food is available?

The likely explanation is competition for something rare that they especially need: a nesting cavity in an open area. The most common sites are widely spaced, having been produced by woodpeckers, which are territorial. The swallows' rush back early in spring gives them an advantage in finding and occupying one of the typically few available cavities, which are crucial to their nesting. Although this makes perfect sense, my study site at the clearing in the woods raised a significant question. I had put up nine nest-boxes. There should have been no competition for a nest site, yet to my surprise I had seen the swallows fight viciously. Would this behavior be repeated?

It was by now clear that white feathers weren't the only mystery related to the behavior of tree swallows. I realized that I had to look at everything and anything, accepting that more unknowns than answers might be uncovered. Everything is connected, and nothing can be excluded, especially when you don't know what the hell is going on.

THE SWALLOWS RETURNED to the clearing on April 15, three weeks earlier than their arrival the past year, and this time they showed up together. I heard their twittering at 6:35 a.m., and the moment they came, they perched side by side, like a long-attached couple. They settled onto the same bare limb of the black locust tree at the edge of the clearing, as they had done for the past two years. A further sign of their being back home: within seconds of landing one of the pair swooped down from that high perch to land directly in the entrance of the same nest-box (there were eight others available) used last year. It was next to the garden, on a pole two meters from the ground. The second bird followed, and both entered the nest-box without hesitation.

The entrance to this box has a small visorlike roof overhang, which would not have been visible from their high perch. Their direct flight to it indicated their familiarity with their home, virtually confirming it was *the* pair from last year. They had therefore likely considered their previous nesting here a success. I had removed the debris from their previous nest, to give them a fresh start. However, despite the early return, this year the nesting would turn out to be much delayed, and that would have serious consequences.

Throughout the next month, long before they started to build their nest, the pair came every dawn to perch on "their" specific twig on the locust tree, the only tree they used, of all the innumerable trees growing in and around the clearing. Whenever another swallow appeared on the horizon, the pair went into a frenzy, vocalizing with vigor. One then swooped down to the nest-box to perch either on it or in its entrance, as if to block entry, while the second rose into the air and pursued the intruder. Chases against singles and pairs occurred daily, though none resulted in the physical grappling matches I had observed the year before. I still did not know which swallow was doing what, since from a distance I could not distinguish which was the male and which the female.

Finally, on May 4, the female brought a single piece of grass into the nest-box — a sign that nest building had begun. The pair had plenty of time, however, to work on the nest before incubating; upon arriving at the clearing at dawn, they sometimes perched idly on a dead twig on the black locust tree for hours at a time. They left it only to chase the daily intruders: other swallows. Not until May 7, when temperatures soared to 81 degrees F and there was bright sunshine, did the female suddenly begin her nest building in earnest. She circled low off the perch, landed on the ground that so far showed not a trace of green, and picked up one frond after another of dried grass, directly in front of the cabin doorstep. After taking one in her bill, she first flew to and past her mate up on the black locust perch, and then she shot down directly into the box. She knew what she was doing from the start, and she worked almost nonstop, finishing a nest of fine grasses in two days. I expected he or she would then go looking for their preferred white feathers to line their nest. Might experiments show who was interested in what?

I HAD SAVED white feathers for the pair. I began offering them after the nest looked finished, and they took each one almost off my fingers when I tossed it into the air. I offered a dark feather, which had white markings, while the pair was perched side by side fifty meters from me, on "their" locust tree. As I stepped out the cabin door and tossed this feather into the air, the reaction took me by surprise — the female instantly dived off her perch, caught it in midair, flew back up past her mate, and carried the feather into the nest, which at that point had no feathers in it. Previously I had thought that it was the male who got the feathers. Wrong again. Maybe it had seemed that way because the female had not found any feathers? Maybe I was able to observe the male's actions more easily because he was less shy of me? I would not resolve this conundrum for a long time, and for now, I was most puzzled about how, surrounded by endless forest, these birds managed to find any feathers at all. I looked but seldom saw one anywhere, and if I did manage to find one, it was rarely white; yet the swallows without fail eventually brought many white or light-colored feathers into the nest.

Not having anticipated the possibility of doing these tests, I had no more white or partially white feathers handy. I then considered using strips of toilet paper as a possible substitute. To my huge surprise, the female swallow accepted one piece of TP after another, catching each before it hit the ground, and if she did not catch it on the first swoop, she circled back and got it on the next. Her mate never once carried anything into or over to the nest, but he was not uninterested — whenever she was busy with nest building, whether bringing grass or TP, he sang sweetly to her in a soft gurgling twitter, his song.

My improvised experiment contradicted my previous observation that only the male brings the feathers. Last year the *male* had repeatedly brought white feathers to the nest, but he had not actually entered the nest-box with them. Instead he handed them off to his mate at the nest entrance, or dropped them there for her to pick up and bring in. Now he didn't budge from his perch. The feathers were apparently not a male nuptial gift, as might be supposed as one of several plausible hypotheses. Many more observations, in other circumstances and with other swallows, would narrow the range of possibilities.

• • •

BUT SOON THE OPPORTUNITIES for making observations on feather choice came to an end. Nest building halted abruptly when cold, rainy weather blew in on May 11 and kept on without letup for three days. The pair was absent until the morning of May 14. Upon their return they immediately mated at the spot where they always perched, their branch on the black locust tree, and after that, they did it again. And it was also the first thing they did at dawn when they arrived the next day.

The female laid her first egg on May 19. At this time the weather went from bad to worse — days and nights of northeast gales rocked the cabin, and constant downpours pelted it. On the 25th, when the swallow laid her fifth egg, it snowed. I had earlier found a pair of red-breasted nuthatches hammering a nest cavity into a rotting spruce stump in the nearby forest, and by now they should already have large nestlings. But when I checked the nest site, there were no adults near it, and none came, and so I finally opened the nest cavity. Six half-grown young lay dead in the soggy nest. Swallows are much more dependent than nuthatches on sunny days because they require flying insects to feed their young. But even the young nuthatches, whose parents glean trees for spiders and sedentary insects to feed them, had starved.

Pelting rain in near-freezing temperatures continued the next day. I was not surprised when the pair of swallows again failed to show up, despite by then having a clutch of five eggs. Would the female return to lay a sixth? It was surprising they had made it to five, since the resources that supported the production of those eggs had not been flying for weeks, and each egg, weighing 1.8 grams, is a significant mass in relation to a bird's body weight of 20 grams. To lay the eggs, the female may have depleted her energy resources to the edge of starvation; for days and nights the wind had whipped the trees, and cold downpours continued ceaselessly, preventing her from feeding on insects in the air.

For the female, the long hours spent guarding from her perch on the black locust tree, chasing others off every day, building a nest, and producing a clutch of eggs when there was little food to support metabolic function, had likely drained the reserves. The last days of horrendous weather may have been "the last straw" for an otherwise valiant effort.

A GORGEOUSLY CLEAR, still day finally dawned on May 27. The swallows had been missing for nearly a week, and I could not imagine how they had coped and where they had been during the rainy and foodless days. But already before sunup, at 5:50 a.m., the pair returned. But as I watched them, something seemed different. I could not at first pinpoint it. Then something curious happened: the male of the just-arrived pair mounted the female and mated, in consecutive bouts of six, five, five more, four, and then two more times, and then still one more mating. After these twenty-three matings, he took a short break before performing two more bouts, of five and then eight times. The bouts were separated by a short flight over and around the female's back.

Frequent mating is the tree swallow's main claim to fame. There are numerous scientific publications devoted to the topic, especially the "extra pair" phenomenon (matings with other than the usual partner) and sperm competition. However, *after* the eggs have been laid, there is nothing to be accomplished by more mating. It is then a wasted effort as well as a distraction that may put the birds at risk for predation. Since the pair in the clearing had *already* mated frequently before their five eggs were laid, this behavior was a clue that I was observing either an entirely new pair or, more likely, a new female. Had the previous male returned and found his mate gone, and the new female here instead? Other swallows over previous weeks (since April 15, or forty-two days earlier) had been trying to get into this territory and use the same nest-box. If new, this female, unburdened by a clutch of eggs, had survived the storms and now had her chance for a nest.

AFTER THE SWALLOWS had mated on the beautifully clear but frosty morning of May 27, another clue hinted that the female was not the same one that had laid the five eggs: this female only *looked* into the box with the five eggs from the entrance; she did not enter.

The pair then left after flying a couple of loops around the clearing, but at 11:50 a.m. the female again flew to the nest-box and continued to act strange — perching five successive times at the nest entrance, each time hanging there, repeatedly ducking her head in, but not venturing farther. Why should she hesitate like this if she was the female that had laid the eggs? The disappearance of the first female was not unusual — birds routinely leave the first eggs of their clutch without incubating them, sometimes for over two weeks (as do ducks with a large clutch). In order for chicks to hatch synchronously, they have to develop synchronously. This means that incubation can begin only after the full clutch has been laid, one egg at a time. There was no reason to suppose these swallow eggs were no longer fertile. But fertile or infertile, that would not cause this female's cautious approach to them. The male swallow, meanwhile, roosted not on the usual perch on the black locust, but rather on top of a pole near the nest-box. When the female left the box and circled the clearing, he also commenced peeking into the nest-box entrance, staying perched there for two full minutes. Like her, he repeatedly only peeked in. Therefore, it seemed that both swallows were likely newcomers here.

The next morning, the 28th, as gorgeous as the one before, I got up as usual before daylight to try to beat the swallows to the clearing. But on this day the pair did not arrive until 7:30 a.m. The female peeked into the nest-box at least a dozen times in succession, but again, she never slipped in. In the weeks before, both swallows at the box had routinely practically dived in; neither had ever shown any hesitation. Then I had another thought. Might this be a pair that had nested here previously, and now hesitated to enter the nest-box because something seemed wrong inside?

I checked. No predator was lurking in the nest-box. Nothing had changed. It still held five eggs on toilet paper, with only two white feathers added. Presumably this pair was now, with some caution, going to take over the nest. They might be new here, but again, a detail of behavior suggested otherwise: their mating. The fact of the swallows' mating thirty-six times in the first couple of hours here, and their uncontested residency, seemed very unusual — *unless* they or at least the male *already knew* of the nesting possibilities here. They may have been a pair that this year's residents had battled.

Not expecting anything exciting the next day, I treated myself to staying in bed until after dawn. As I got up, then built the fire to heat coffee, I all the while glanced out the windows at the nest-boxes, but I saw no birds. Just to be sure one had not slipped into the box, deposited a sixth egg, and then left, I went out, lifted the door to the box, and took a peek — still just five eggs on toilet paper, as on the past three days. Nothing of note, except that a goldfinch flitted by, and I had not seen any all winter. I wrote up some general notes about the hummingbirds now at the apple blossoms, a pair of blue jays, a chestnut-sided warbler singing on his usual perch, and an American robin on three eggs in her nest, on a log at the side of the cabin.

It was not till 8:30 a.m. that I heard swallows coming, and they circled the clearing. The male then dived down and perched on the nest-box with the five eggs. He glistened like a polished blue-green jewel in the sunlight, as the female continued circling. Finally, she swerved over to another nest-box — one I had never seen the swallows visit. She fluttered at its entrance, then landed there and spent two minutes ducking her head in and out, as she had done on the previous day at the nest-box with the five eggs. But this new box was empty.

The male started his sweet liquid gurgling song of approval as she ducked deeper into the empty box. He then flew over to her and perched on the pole holding the box. When she momentarily flew onto the top of the box, he fluttered briefly over her back in what looked to me like a mock mating, with no contact. She then immediately returned to the same box entrance, and for minutes she partially slipped in and out, and then went all the way in. After a minute he joined her and also entered the new box. I believed now that they would accept this box — it showed no evidence of an owner.

The pair stayed five minutes inside the new nest-box before they came out, one behind the other, circled, returned to the box and post, and then mated four times. They then both circled around the clearing until he landed near me, and I heard his soft

gurgling song, which you have to be close to, to notice. It's meant
for private ears — hers.

The pair of swallows now took turns, one flying around and
the other spending time inside or on top of the newly inspected
empty box. They slipped in and out without hesitation, and their
soft voices oozed contentment — until the male, perched on the
box, erupted in a fit of excited loud chittering. But instead of
flying upward, he only looked up, and as I followed his line of
vision, I saw a third swallow approach. His mate flew up to it,
but there were no mad chases as before, when an intruder came.
Instead, the female of the pair was flying *with* this third bird,
and she was not in the least bit aggressive toward it. Before, the
female had been the main chaser while the male of the couple
perched and defended the nest-box entrance. This time, however, the male was highly agitated whenever these two birds
came near him on the nest-box, and each time he flew to block
the entrance. Unlike what I was used to seeing, namely, his flying
off to give chase, he defended the nest-box itself rather than the
airspace above it, where the female continued to interact seemingly playfully with the visitor. These two made several circuits
before the visitor left, and the female swallow then rejoined her
mate at the nest-box.

What happened next surprised me: for a second or two after
the female landed beside the male, he fluffed himself out, lowered his head at her, and clicked his bill, producing a snapping
sound (which swallows do during the final centimeters of their
attack dives to the back of a person's head — or at least mine).
This behavior indicates irritation or disapproval, and in this case
— was it a sign of disapproval? Her joining the intruder must
have been irritating to this male swallow, because he had been

powerless to give chase while he was guarding the nest-box entrance. But his flash of emotion was brief: a few seconds after she went into the box, he again commenced his sweet gurgling song. There was now, it seemed, much more going on than I had expected, and more than could be documented, but none of it could be dismissed as irrelevant.

Although this second late pair of swallows had saved energy resources (such as stored fat) until now, I still wondered if the female of the first pair might possibly still be alive and return later. And on May 29 at dawn, under a hazy sky turning to light rain, the single female swallow that piqued this question did come. She stayed until 9:15 a.m., on the black locust tree, the favorite perch of the previous pair.

The next dawn I vaulted out of bed at 4:58 a.m. because it looked like a bright sunny day was in the offing. I wondered if the other pair, which had always arrived early, should visit again. And at 5:34 a.m. a single swallow once again arrived, circled high, and chittered continuously, so that I could follow it by sound alone. It continued for minutes on end before landing on the same branch of the black locust where the single swallow had landed yesterday. It stayed a couple of minutes, circled the clearing for a few more, and then returned to the same spot before circling again.

MY NOTES WERE SPARSE for the next two weeks, in part because of the absence of all swallows during still another bout of bad weather. But by June 15 the most recent pair had built their grass nest in the second box. So far there were no feathers in it. The pair had mated, with egg laying about to begin. Nest parasitization by "egg dumping," a well-known "alternate reproductive strategy" of birds, could now be a possibility at their nest. But how? Any egg laid into another's nest before the female nest owner had laid one of her own should be seen by both the nest owners as not belonging to them. But a parasite egg deposited after a clutch had been laid would likely be accepted by the swallows — their immaculate white eggs are visually indistinguishable one from another. Any swallow hoping to achieve a successful egg insertion would have to get the timing right. If such a maneuver was to be attempted at this nest, it should happen about now. It was time for me to be super-alert in order to see it. I was especially interested in learning whether the white feathers might have some function in averting such an egg insertion.

The next day, June 16, I awoke at dawn, hearing a loud continuous chirping. Expecting that something unusual was in progress, I instantly jumped out of bed and ran out to see a swallow flying so high, it was almost out of sight. Its chirping was ceaseless, occurring several times per second. I followed the swallow by ear for at least twenty minutes, and then by eye for ten minutes more, when at 5:20 a.m. it finally dived down and, in a smooth swoop, slipped into the nest-box of the pair I had most recently been monitoring. It was the female, but the male followed and landed on top of the nest-box, where he then sat silent for the next twenty-seven minutes, swiveling his head several times per second, in all directions, all of that time. By 5:57 a.m., he was still silent but flew up, and I followed him by eye for the next nine minutes before he made a soft chirp and flew over the nest-box, and then resumed perching on it. I suspected she was still inside, laying an egg. There was so far not a single white feather lining the nest. This late in the season, with the grass grown high, it might be impossible to find any feathers, and I was tempted to again toss TP into the air but decided against it, having already tested it once.

The male swallow continued to perch on the box until 6:26 a.m., when he finally fluttered down to look into the entrance. He then flew off to circle around silently for a couple of minutes before entering the box. He stayed in for only a minute before coming back out and perching on the locust tree, which he did until 6:37 a.m. During that time another pair arrived, and he flew to intercept them and tussled with one in a fight, after which the visiting pair left. The male again entered the nest-box briefly, then perched, silent, on the locust tree. The female swallow meanwhile had laid the first egg.

THE BEGINNING OF EGG LAYING IS, for most birds, the end of nest building. With usually one egg added per day until the clutch is complete and incubation starts, most birds are secretive. They slip into the nest to lay the egg, but then they stay away. Reduced activity near the nest decreases the chances of disclosing its location to predators — the survival of the nest relies on its being hidden. But the tree swallows I was observing had conspicuous nest locations — in an open clearing, in a box created expressly for nesting. This means that they had to stay close by to guard the nest, keeping intruders at bay, possibly to the point of mounting a physical defense.

The situation is somewhat different for tree swallows with nests in natural conditions. Such nests are safely ensconced in hollow trees, where there is usually small danger of predation. But even in that environment, during egg laying the tree swallows demonstrate a high level of nest guarding. This behavior, however, is not intended to protect the nest from predators. It is directed against others of their own kind — female swallows trying to dump eggs into others' nests. Females likely do this because they have no nesting place of their own.

When a female succeeds in foisting parental care on unsuspecting victims, the cost to the nest can be high. There will be too many babies, and some will die. We have seen this in Chapter 2. Such a situation would have instigated an evolutionary arms race between swallows guarding a nest and swallows aiming to sneak an egg into such a nest — means of inserting an egg versus means of averting this maneuver. Cuckoos and cowbirds pull off egg dumping routinely under more difficult situations, namely, into the nests of other species of birds, which should and often do recognize a foreign egg by its contrasting coloration. Indeed, that is likely one of the main reasons for the evolution of different egg colorations in the first place. In cuckoo species this well-known arms race is exerted through (but not restricted to) egg-color matching of parasite eggs to host eggs, the host's recognition of mismatched eggs, and a variety of behaviors to deal with such. But within the same species, it should be easier for perpetrators to accomplish nest parasitism because egg matching will be automatically almost perfect. Therefore, potent counterstrategies are expected.

Timing is extremely critical: an egg that is inserted a day or two late will become the runt of the litter (except among birds with precocial young, which are born with the ability to feed themselves from the start), prone to starve in the competition for food among the young. Conversely, an egg inserted before the host has laid one of her own could potentially be discriminated against, so long as the host knows whether she has laid an egg. After she has laid an egg and a foreign one is inserted, the chances are even that any egg she removes or destroys will be her own. Thus, for the egg cuckold to be successful, the female should insert the parasite sometime during the time of egg laying, and assess the nest contents to time it right. Egg ovulation is hormonally triggered from sensory stimulation through the nervous system, and it would take another twenty-four hours or so to produce an egg ready for laying. And so I wondered if the white feathers that tree swallows place in their nests are an anti-cuckolding strategy to scramble the signal, the time when eggs are going to be laid.

WITH EGG LAYING started and not a single feather in the nest, would the pair of swallows now put feathers in, to help hide the eggs and shield them from sight? A female who intends to dump an egg into another's nest would need to monitor the nest contents, in order to get her timing right. With large white feathers covering the eggs, she would have no idea whether the nest contained no eggs, a clutch of eggs, or small young. So for the resident swallows *not* to apply feathers before or when the eggs are being laid would seem counterproductive; without such cover, the eggs would look as obvious as lights in the dark, making cuckolding far easier.

I was now interested in keeping a close watch on the nest and the sky, to see if I could reject my first hypothesis, namely, that the swallows had, due to the lateness of the season, made do without a nest lining simply because there were no feathers around and they didn't have time to look for them before egg laying began. The next few days could be important. So, after 7:35 a.m., when both birds were gone and the first egg lay on the bare grass in the nest, I kept watching: this could be a prime moment for egg dumping. However, a disaster that can confidently be anticipated is precisely the one that is most strongly guarded against, and hence the disaster becomes rare. Airport security is a good example. Planes practically never blow up from bombs inserted by passengers, yet costly and time-consuming security measures exist despite the virtual empirical day-to-day absence of the feared danger.

As I was sitting relaxed on a woodpile in the sunshine an hour later, the pair of swallows returned. The female slipped into the nest-box (now containing the first egg) without hesitation, and the male came right behind her, fluttered briefly, and then perched on top of the box. She emerged after a minute or two and circled low over the clearing, then reentered the box. After a minute or so she exited and flew to the precise spot on the branch of the black locust where the pair had mated. As she landed, he took off from the top of the box and fluttered over her. They mated (six times) again.

A half-hour later another pair arrived, and one of this visiting pair also flew to the nest entrance of the first pair's box. Clearly, this second pair had not, at this late date for tree swallow nesting, yet found a nesting site. They were far behind schedule but still had interest in nesting. One bird hung at the nest-box entrance long enough for me to see it was a female. The nest-owner male, which had in the meantime entered the nest-box, shot out and vigorously chased this bold female intruder, all the while excitedly chittering and screeching. It was a brief chase; the intruding pair beat a hasty retreat.

The mere fact that a second pair was still appraising a nest-box suggested that suitable nest sites were in demand, hence rare. It puzzled me that the eight other potentially available nest-boxes I had provided were not used. After all, tree swallows are well known to nest near each other. I now had the hunch that the swallows were protecting *sky space* rather than one nest-box or another.

The defending male returned and resumed his perch on top of the nest-box, while his mate continued to fly low, landing a couple of times on the ground, where she picked up tiny pieces of grass. But this grass picking was only a gesture now; she dropped what she had picked up before flying back into the nest. It was not grass that was needed now, and as if he understood, the male then left his perch on the box and disappeared into the distance.

In my ten spot checks of the nest until 4 p.m., when it started to rain, every time at least one of the birds was nearby. The pair maintained a presence close to the nest nearly constantly. No new egg had appeared in the newly finished grass nest; it still contained only the one egg laid in the morning.

On this first day of egg laying, as on all subsequent ones, the swallows maintained a conspicuous nest-guarding presence and vigorous responses against other tree swallows.

ON JUNE 17, the second day of egg laying for the pair, I again heard nonstop predawn chirping at 5 a.m., and rushed out to see a swallow fluttering high in the sky. Visually it could have been mistaken for a bat, but its loud chirps, about three per second, proved it was a swallow. It was at times too distant for me to follow visually, but for a half-hour, until 5:30 a.m., it took no break in its chirping and circling. It was the male: his flight was less fluttering than the female's. His fluttering was interrupted by glides, giving him a calm demeanor. To confirm, I checked the nest: indeed, the female was inside. Like yesterday, she remained mostly silent, making only a very soft, almost whispered chirp whenever her mate flew low over the box.

He continued his sky dance in the same way for another hour and a quarter, until 6:15 a.m., when he came down to perch on the locust tree. Eight minutes later, to my great surprise, another swallow (of which gender I could not say) arrived. The male swallow joined it, and they flew out of sight. I presumed at first that the companion could have been his mate, and I had missed seeing her exit the box. So I immediately checked the box, and there she was, still inside! The male had left with another, showing no antagonism toward it. But he returned quickly and, unexpectedly, he dive-bombed me — the first time he did such a thing. He then made soft cooing calls in several successive passes over the nest-box, apparently letting his mate know that he was back, and everything was OK. And then he settled on his perch on the locust tree.

Finally, at 6:46 a.m., he landed on the box, peeked in, and flew back up to perch quietly on the locust tree until 7:12 a.m. Then he began lifting one wing after the other, stretching them, scratching the back of his head with a foot, swiveling his head, and making sweet-sounding chirps. Then he suddenly took off into the distance, where he was again joined by another swallow. Was it again a visitor, or had I by watching the male on the tree missed seeing his mate leave the nest? I ran to the box to check; this time she was not inside, but the nest now held a second egg. The male swallow returned in a half-hour and perched at the nest. I checked six more times over the rest of the day, and always saw one swallow near, and one interaction among three.

Since most birds stay off the nest unless they are in the process of depositing the egg, and do not stay on the eggs until incubation begins, overnighting in the nest would be superfluous — unless it might be a good method to guard against early-bird egg dumpers? Local sunset now was at 7:26 p.m., and so I went to check the box at 8 p.m. I heard the male's alarm calls, and he whooshed at me and came within ten centimeters of my head. I immediately withdrew, and he became silent again, continuing to circle and making a whispering chirp whenever passing the box. He continued to patrol near the nest even as it started to get dark. At 8:11 p.m. his mate joined him. She dived into the box, and he followed right behind her. That could be the end of the day, I thought, with both of them overnighting together on the first two eggs. But just in case, I waited and watched a little longer, and within one minute the male came back out of the nestbox, to again circle silently; not until 8:21 p.m. did he disappear high in the night sky toward the east. I waited to check the box after another eight minutes. Darkness had fallen. She had stayed inside, and he did not return.

I do not know how she desisted from incubation while staying in the nest overnight, but if she were incubating, then the spread of hatching dates would be a week apart. It wasn't. I suspected, therefore, that she either lowered her body temperature at night, thereby not incubating and thus also saving energy reserves, or she did not allow her blood-engorged brood patch to make contact with the eggs.

The next morning, June 18, at 4:56 a.m., an hour after sunrise, again the male swallow chirped while flying high and then came closer, while the female was in the box. He circled and vocalized overhead nonstop until 7 a.m., when he became silent, except for an occasional whispering call when he came near the box, entered it, came back out within thirty seconds, and then left. When I checked the box a few minutes later, it contained three eggs and also the first feather. It was black. I would have bet money on its being white! I felt taunted.

At 2 p.m. one swallow was peeking out the nest hole, but after ten minutes it ducked back down, and then a few minutes later one flew off. I ran to the box to check: no bird was inside, but no new egg had been added. There were now two white feathers in addition to the one black, and an hour later, still another. That evening at 8:20 p.m. the female was again on the nest, but the male had left. He returned the next morning, June 19, at 5 a.m., and as usual he chirped as he arrived from a distance. As before, he also continued circling high over the clearing until 5:25 a.m., when he stopped chirping but continued circling in silence, except for occasional soft vocalizations when he swung by the nest-box. At 5:50 a.m. I suddenly saw two swallows flying low, and the nest contained the fourth egg, and the fourth white feather.

June 20 promised to be a beautiful day, just like the one before. As usual by now, I heard the first swallow chirp in the sky at 4:57 a.m., but within a few minutes the chirps became infrequent. The female left the nest at 5:33 a.m., and as she flew her first loop, I saw her release a large fecal load. She then immediately returned to the nest-box, to lay her fifth, and last, egg. Both birds left after she had laid it, and in their brief absence I retrieved one of the five eggs from the TP-lined nest that the *first* pair had deserted, painted spots of red nail polish on it (so that I and possibly the swallows could identify it), and added it to this newer clutch, to find out if the pair would accept it. One more white feather had been added to the nest. Might a *spotted* egg of the same kind be recognized as different, and would or could the parent swallows discriminate against it, and if so, how?

In contrast with all the previous days, after the clutch was finished on June 20, and I had added the extra egg (the sixth), the nest had ten white feathers, and by the time the young hatched, on July 4–6, thirty-five more feathers had been added. Eighteen were white, thirteen mostly white, six brown, seven with light fuzz at one end, and one black. Also included in the nest was one Band-Aid, with a white pad. Given the unbroken surrounding forest and the unlikely chance of a Band-Aid being found in a tree, I strongly suspected the swallows had been to the public beach at the lake seven kilometers away.

After incubation had started, no more swallows visited the clearing, there was no nest guarding, and I did not see or hear any more chases.

The spotted swallow egg remained in the nest but did not hatch, perhaps because it was by then not fresh enough, or perhaps because of something poisonous in the red nail polish. Alternatively, perhaps the birds had shoved it to the side during incubation. Several years previously, I had observed a female swallow that had apparently done the same with the parasite chickadee eggs deposited in a swallow nest.

The weather continued to be cold and wet, and July brought many days of continuously pounding rain. As was the case in 2011, this clutch faced death by starvation, only even more cruelly. When I examined the nest-box, I was saddened but not surprised to find the five young dead. There was no sign of injury. The reason for their death was obvious.

There were huge differences in size among these young, even

though they had hatched within a day of one another. In the life-and-death contest to get the limited food being delivered, one or another of the young got a slight edge that later translated into a sizable advantage in getting the incoming food. The two smallest were trodden down flat in the guano on the nest bottom, and another two of the dead faced the nest entrance. My insertion of the additional egg had, I was glad to learn, no effect on this outcome, since it did not hatch. This graphic example of the great cost of having too many young showed how essential it was for the parents to invest great effort in defending against egg dumping. One extra egg could endanger an entire clutch. When tree swallows had again nested in the same box this year, all had seemed "normal." But, as Thoreau penned in his journals, "Facts fall from the poetic observer as ripe seeds." The fact that I could deposit an extra swallow's egg suggested that an extra-pair *swallow* could have done the same with a fertile egg.

"My" tree swallows were beginning to reveal their struggles in a way that felt far richer and more exciting than the studies of trees swallows that I was reading. Researchers had studied swallows living in identical boxes set up in huge symmetrical grids in order to get consistent and statistically significant results. I was now eager to see the swallows the next spring, to observe how individuals might deal with the more natural situation of this clearing in the woods.

5

NESTING NOVICE — 2014

On April 11, 2014, the remaining snow was melting in the clearing, although it was still three feet deep in the woods. Despite the rising air temperature, not a single leaf bud had yet opened. But birds were racing back to their old territories or establishing new ones. Geese, ducks, and red-winged blackbirds had already returned to the marshes. But the tree swallows had not yet arrived. I could not have missed them — they are demonstrative enough to be noticed when they come. Most birds tend not to return to a place where they have experienced repeated failure, and the nestings of the two pairs of swallows the previous year had failed. I had been wondering if they would be back this year. The answer came at 9:20 a.m. on this windy and cloudy morning.

One swallow flew alone from the direction of Webb Lake, where swallows had gathered in the past. It would have skimmed over the water and the adjacent fields and wetlands before coming here. It chirped as it came sailing into the clearing. Hundreds of trees surround the clearing, yet this swallow chose to land within ten centimeters of where the swallows had "always" perched — the locust tree — and there almost immediately it started to preen. It groomed itself for about three minutes, then flew down to the nest-box where the babies had died of starvation last spring. It peeked in for only about a second before it flew up again, circled for a minute, and then returned to its perch on the locust tree, where it stayed only a few minutes, chirping weakly and then vanishing in the eastern sky. By 10 a.m. it had returned again (to the same perch) but this time stayed ten minutes before leaving, again returning in a half-hour and staying six minutes, and again leaving and then returning three times over the next three hours.

Throughout its trips, the swallow had been alone, and it always perched within a foot of the same spot on that tree. After visiting the last-used nest-box, it visited three more, spending only several seconds at each entrance to peek in; then it flew back up to its perch or to another box. Except for occasionally making a soft, barely audible cheep, it was silent.

The swallow returned at 6:55 a.m. the next day, when the temperature was only a single degree above freezing. With the woods still covered with meter-deep snow, few if any flying insects were present. How had this bird managed to survive here so early in the season? As on the day before, it had come alone but did not recheck the nest-boxes. It again made a few muted cheeps and left after less than a minute, returning twice more before noon. Each time it used the same perches and stayed only a few minutes.

• • •

THEN IT CAME NO MORE. I waited day after day for weeks. It must have been a bird returning home after a year, but its mate, and potentially another bird familiar with this nest site from a previous year, had not yet returned. During that swallowless time, "the" phoebe pair rebuilt their nest under the eaves of the cabin, a male yellow-bellied sapsucker drummed on an aged apple tree in the clearing, a male woodcock sky-danced. A crowd of hundreds of dark-eyed juncos, along with one white-throated sparrow, dropped in from their unfinished migrations and stayed two days, feeding on our black sunflower seeds. Purple finches, evening grosbeaks, American goldfinches, and mourning doves came to our feeders daily. A golden-crowned kinglet stopped by, and also an eastern bluebird. The robin sang at almost any time of the day, and the barred owl hooted every night. A song sparrow stayed for a day, and a northern flicker checked out the previously used nest hole in the cabin wall, but it did not stay. As in a previous year, three male red-winged blackbirds flew in, landed on the maples next to the cabin, and left within five minutes. Three common grackles passed over, and blue-headed vireos daily poured out their languid songs in the surrounding woods, as winter wrens joined in with their vibrant refrains. But as April slipped into May, there was still not one more sound or sighting of a tree swallow. Not until May 6, three weeks after my last sighting of the lone laconic returnee, did one show up.

It arrived alone at 5:10 a.m., but then suddenly, at 7:30 a.m., a pair of tree swallows were swirling all over the clearing, chirping loudly, nonstop. They continued their conspicuous display, showing no sign of either wanting to land or leave, as though they owned the patch of sky over the clearing. They didn't pay much attention to the nest-boxes, but fluttered briefly and chittered loudly at one that was tilted and had a small knothole at one side, which should have been visible to a bird circling the clearing. The hole was too small to enter, and the two flew off into the distance.

A pair of swallows came shortly after dawn the next morning and swirled around the clearing, as on the day before. These two never landed on the locust tree. Several times they fluttered in front of a nest-box, then landed on it. The second swallow, a male, then flew down and perched on a nearby post, where he stayed several minutes, making a sweet twittering song; he then flew up to his mate on the box, ducked his head into the entrance several times, then slipped inside and stayed there a full minute, apparently pleased with the place. I returned to our cabin to celebrate with hot coffee and breakfast, confident that this pair, having found a nest site, would remain to nest here.

When I went back out at 7:15 a.m. on this beautiful morning, a swallow was indeed perched on what I had already presumed would be, or already was, the pair's chosen box. Given its bright sheen, this bird was surely the male. Very slowly, step by slow step, and with many stops, I approached him, realizing he and his mate were new here and would not know me. It did not yet know what might or might not seem threatening about these surroundings. So, wanting to tame the pair, I needed to start immediately; if they noticed me later, *after* they knew every bush, tree, and stump, I, a human, would be something out of the ordinary, and hence potentially alarming. I intended to teach the swallows that I was like any other feature of this landscape, like the post, the solar panel, or the apple tree. I would then become invisible to them, and their activities would be unaffected by my presence.

Almost from the beginning, the male allowed me to approach within two to three meters, where he perched near me for thirty-five minutes before taking wing and then coming back to slip into the box. He stayed inside for fifteen minutes while I walked over to him and again stood next to him as he peeked out. I edged closer then, to a meter away, and still he seemed not to notice me. After finally exiting the nest-box, he circled around the clearing and then examined a second box. He again chirped softly, even after entering it. He stayed inside a minute before coming back out to circle and then leave. He returned a half-hour later, this time with the female (her duller plumage made her easy to identify). He again entered the last-visited box, but she showed no apparent interest in it.

Neither of this pair ever perched on the black locust tree that the previous pair had used exclusively. This male instead routinely used a specific garden post, one of the fifteen that ringed our garden, as a perch. The female spent most of her time on the wing over the clearing, but on the morning of May 8 she examined four of the nest-boxes. Given so much choice, how would the two swallows reach a consensus?

She was often absent for long durations, but whenever she returned, swirling around the clearing, he responded instantly to her presence by flying to either one or the other of eventually only two boxes, hanging, head up, in front of an entrance and making loud chittering calls. Whenever she came close to him, he ducked down into this box, casting his vote. She assented by landing at the entrance and slipping in to join him. When he came back out, she followed, circled, and then left, as he returned to his post. This sequence was repeated, it appeared, for many of the nest-boxes, but soon only between two boxes, until later in the day they came to prefer the one nearest the cabin. It has a front panel that I could detach in a second, to allow a full lateral inside view.

May 9 dawned at 38 degrees F, with two barred owls calling back and forth, a sapsucker's drumming, and a blue-headed vireo's singing. It was not until about 6 a.m. that a single swallow arrived and perched on the black locust tree. It gave a brief soft chortling song, but as soon as the pair arrived, a vigorous chase ensued, until all three were out of sight. The pair returned in minutes. But the male that had been chased, or possibly its mate, soon returned as well. Long before I saw him coming, the pair had already sounded an alert, flew up, chittered loudly, and chased the interloper away.

Scarcely an hour later, another single swallow arrived. The male of the pair vocalized as before, but this time instead of intercepting the stranger in the air, he flew off his post down to the pair's chosen nest-box and perched at the entrance, guarding it. *This* interloper, given its drab colors, was a female, or a second-year juvenile mimicking one. Tree swallows are the only swallows that retain their drab juvenile garb for two years, a strategy thought to reduce territorial aggression (which is restricted mainly to adult birds in breeding plumage). Yet in moments, a vicious fight erupted. The two females met in midair, grappled, and fluttered onto the ground in a ball of feathers. There they disentangled themselves, flew up, and repeated the same performance. After still more chases, the newly arrived bird left the clearing and the resident one returned to her mate, who had meanwhile guarded the nest-box entrance by perching in it and facing out. He now joined her, and the pair left to chase the intruder; all three vanished as specks into the distance. After the pair returned from their wild chase, the female perched at the nest-box entrance, while the male returned to his perch on the pole. There he sang his sweet chortling notes, one after the other, each of about seven syllables lasting two seconds, with the last syllables so short, they almost ran together.

His melodic, tinkling, gurgling song was not like the territorial advertisement of most songbirds that nest here in these woods, proclaiming ownership and warning others to stay away. It was at times almost a whisper, perhaps a signal indicating approval, in this case of a highly desired potential nest site meant strictly for the specific female close by. Like other male swallows, he sang only when *she* was there, which likely helped keep her committed to that nest site and encouraged nest building.

Over the next two days, May 10–11, there was an almost continual repetition of the same behaviors, involving apparently two extra-pair females, one of which tangled in physical combat with the resident female; the other by contrast sometimes peacefully accompanied the pair. Once, two females perched simultaneously at the entrance of the box, then fought briefly, while the male was inside. My notes became a jumble as I tried to get more and more details in order and sort out what the male and at least two females were up to. So far not a single blade of grass had been brought into the nest-box.

ON MAY 13 a pair of swallows came and generated a new round of vigorous and vociferous chases. But the antagonism this time was not restricted to, and didn't end in, the clearing. It continued in the air, far into the distance, where two swallows engaged in a physical tussle until, spiraling, they fell from the sky. This was also the first day the resident pair showed a hint of nest-making activity, and it was the male who lifted the first straw, literally. He flew to the ground, picked up a piece of dry grass, and carried it into the nest-box while the female perched nearby, on the solar panel. By noon he had made three more straw-carrying trips to the box. She had not budged. This was strange, I thought, because I had so far seen only females build the nest. However, it soon appeared that by example, he was stimulating her to start the work on the nest. When he flew to the ground, she swooped partway down as though following, and once even landed beside him as he picked up a straw. He also nudged her physically as both perched on the solar panel, and also in flight, and once even at the box entrance.

That afternoon, a second female or yearling juvenile again arrived, as had happened frequently before. But this one was silent and moved about unopposed, entering the nest-box just when both members of the resident pair happened to be inside. All three then stayed quietly together inside the box, and only after about five minutes did one of the females come out, fly around the clearing, and then leave.

Most of the action on this day had occurred directly before my eyes, as I stood at the post where the male perched, an arm's length from me. As a test, I reached up slowly and touched him on the tail; he did not take flight.

The next day (May 14) he again brought several pieces of fine dry grass into the nest-box. It was, as before, an apparent inducement for the female to get busy with nest building, because she again landed on the ground next to him. As he picked up the grass, she pulled at a grass stem, dropped it, and flew, empty-billed, to the box and went in.

Two days later the female of this pair still seemed clueless about what to do, and the male was still "demonstrating" the task to her by performing nest-making behaviors. She again flew to him when he tugged at a piece of grass, and then went back to the box, empty-billed. When he brought a long piece of grass (about fifteen centimeters) and partially dragged it into the nest entrance, she came, took it from him, flew off with it, and dropped it.

Enough time had passed by May 18 for the swallows to have built several nests, but the female still had not "caught on," and the male had not done more than give hints about what to do. The pair now more frequently perched side by side on the solar panel but still stayed about a half-meter apart (unlike last year's pair, which routinely perched four to six centimeters apart on the black locust tree). He still repeatedly flew to the box, perched at the entrance, and quietly sang his sweet song there, while looking back to her over his shoulder. But his apparent efforts to stimulate her seemed so far to have little effect.

The female swallow had not yet ordered the pieces of nest-making behavior into the proper sequence. Either she was a young swallow hatched last year, or the union of this pair had been too short for the hormones that stimulate nest-building behavior to kick in. But this morning I saw change. For the first time she picked up a piece of grass, flew to the box, and perched at the entrance with the grass. However, she then left to return to the solar panel — with the grass still in her bill! The male took note, of either her progress or negligence, because he responded by hovering over her back, making scratchy chittering calls and short jerky flights directly *at* her. It looked as though he was trying to shoo her off, and she then did immediately return to the box with her piece of grass, to again perch at the entrance with it — and then finally took it in. It was the first time she had completed the sequence of delivering nest material into the nest-box. When she came out, she flew high above the clearing. The male left his stump and joined her in flight, and they both left.

THE OBSERVATIONS I RECORDED about this behavior reflect "pure" interest in the swallows as such; that is, they were not colored by any bias toward solving a specific problem or answering a certain question. Yet the more I saw, the more I realized that I had witnessed something amazing. These birds have hardwired instinctive responses, as we do. But the behavior I'd observed strongly suggested that, in addition to instincts, the swallows have intentions and can modify their behavior through learning to accomplish a specific task. Ironically, if nest building was always done correctly and perfectly, the way a bee makes its hexagonal honeycomb of wax, then such intentions would be superfluous — simpler mechanisms had sufficed. What the swallows did reduces the perceived gulf between them and us. However, this female swallow had only sort of caught on, as she proved three hours later when the pair returned.

The two swallows were perched together on the solar panel when the female flew to the ground to pick up grass directly beside me. I was close enough to hear the snap of her bill as she tried to grab the grass, but she apparently missed it, because she flew empty-billed to the nest-box entrance, where she hung briefly and then flew off, as if forgetting why she had come. The male left the panel and also perched at the nest entrance, but he only peeked in, then immediately returned to the panel. She again flew down to the ground; this time she came up with a piece of grass and actually carried it to the nest entrance. But then, as though again not knowing what she had just done, she flew back to perch next to the male, still holding the grass in her bill; then once more she returned to the nest entrance. But this time she ducked her head inside and dropped the grass in. After this apparent breakthrough, she then, as before, immediately flew off into the distance. As before, he followed, but this time they returned in less than a minute, and he immediately flew to the entrance as she landed on the ground. The ground was covered with dead grass, but she came up with nothing in her bill. Nevertheless, she flew to the nest-box entrance to hang there briefly before again flying away. He again followed her.

She became more focused over the next two days (May 19 and 20) but still remained clumsy. Once, she brought a long, stiff piece of grass and held it crosswise at the nest hole. She tried at least ten times to push in with it, each time flying off and coming back to try again, before she eventually, perhaps by random chance, held it at the proper angle and got it in. The male had been inside the box, vocalizing softly. He still had not brought anything into the nest-box since his initial dozen or so demonstrations with small pieces of grass six days earlier. By the end of the next day (May 21) the bottom of the nest-box had only a thin ring of grass on the bottom.

ON MAY 22, the male arrived at 5:25 a.m. By now his usual perch was the solar panel, not the tall stump directly next to it, a former favorite. The female came an hour and a half later, and the moment she appeared in the sky, he flew to the nest-box and sang there in apparent solicitation. But she perched on the black locust tree. A half-hour later she fluttered over the ground several times without landing, then finally grabbed several small pieces of grass and brought them to the nest-box entrance. Still, she did not enter, but instead flew with her grass load high into the sky, while the male remained on the panel. Relative to us, year-old swallows are hugely physically precocious but not likely rational; they likely do nest building for the fun of it, rather than for accomplishing a functionally logical task.

The male left at 7:17 a.m., and at 8 a.m., when both were finally back, he sang from the panel just before flying to and entering the nest-box; the female followed and landed on it. He stayed inside and chittered there for about a minute before coming back out to fly back to his perch. She then brought grass that she had picked up two meters from me, dropped it into the box, and then flew to perch next to him. After this, she completed thirteen more grass-carrying attempts from 8:15 to 8:31 a.m. Nine of them were successful. Throughout all of this, her first flurry of productive activity, the male did not budge from his perch on the panel nearby. He just watched and sang continuously. There were no other swallows around all day.

Over the next days (May 23–24), the pair continued the same activities, except that his singing became reduced to a whisper as she repeated flight after successful flight (about ten trips per fifteen minutes) to deliver grass. She no longer hesitated to enter the box; she was literally diving in. The male had now stopped visiting the box. I felt pleased to see the female carry long grass stems that flowed a body length behind her. Now when she sometimes, perhaps accidentally, dropped something in flight, she did not continue flying to the box; instead, she turned back to retrieve it in midair.

By May 25 the nest had a deep cup of fine grass stems arranged in circles around the bottom of the box. It looked much like the completed nests of most birds that make cup nests. But this was a tree swallow nest, and it did not yet contain a single feather. Although it looked finished, the female still occasionally brought in another piece of grass. Wondering if she might be interested in feathers now, I tossed one into the air when she was on the panel, near me. She flew over to me, investigated the feather as it fell to the ground, but did not pick it up.

Near noon the pair mated; egg laying was close. The next morning (May 27) at 5:30 a.m. they mated again. Another pair of swallows came, and the resident pair chased them off, but this pursuit involved much less fuss than similar events before nest building had begun. This suggested that competition for a mate was less of an issue than competition for a nest site.

On May 30 the nest, consistent with observations at previous nests, contained two typical pearly-white eggs in a deep nest cup of yellowish-brown, dry, year-old grass. The pair of swallows remained in or around the clearing. The female at times flew to the nest entrance, lingered there, and sometimes also entered, while the male remained nearly the whole time perched on the panel, where he chortled softly.

At 5:30 a.m. the next day it was cool (47ºF) under an overcast sky. The phoebe sang at dawn. A sapsucker drummed. The female swallow peeked out of the nest-box as the male was flying high above the clearing, continually twittering. After a few minutes she flew out and landed on the locust tree, and I checked the nest-box. She did not stir from her perch as I opened and closed the panel of the box in front of her. There were still only two eggs. A minute later she entered the box, and at 7 a.m. she again briefly peeked out before she flew out. Her third egg had just been laid. There was no feather in the nest. This was the moment I'd been waiting for: the lining of the nest with specifically white feathers, if indeed they are chosen specifically because white ones are preferred to others.

6

EXPERIMENTS WITH
FEATHERS — 2014

MALE BIRDS MAY START THE NEST BUILDING, BUT TYPI-
cally the female then makes most of the nest herself, to hold,
hide, and shelter the eggs and the young. Only after inserting
the lining does she begin to lay her eggs on it. However, with
the tree swallows I'd observed so far, creating the feather lining
came mostly after the clutch had been deposited. Was that an
individual aberration, a response to particular circumstances, or
the modus vivendi for this species?

It seemed possible that in those earlier cases, feathers had
simply not been available and the egg-laying schedule could not
be delayed. Yet that could not be the sole reason for the absence
of feathers in the current nest because when I had offered the
female feathers, she had not taken them. Since she had specifically investigated them, she had certainly seen them. Perhaps
they were just not the right kind of feathers?

A bird chooses nesting material based on an innate taste pref-

erence determined by natural selection, which produced the behavior to promote successful nesting. Materials used to make
the nest are utilitarian and specific to the purpose. Cliff swallows' use of mud attaches their nests to cliffs. Hummingbirds'
use of spiderwebs holds small bits of lichen together to form a
cup nest and attach it to vegetation. Tree-nesting herons need
long, strong twigs to bridge branches and create a platform. It is
logically difficult to assign a specific function to white feathers.
But tree swallows had specifically chosen them, along with white
toilet paper, so what precisely did these materials have in common that gave the signal to express the feather-gathering behavior? Was it color alone?

Experiments related to choice can empirically demonstrate
a value. For example, if a child picked up a red apple instead
of a brown one, we would assign the choice to taste, a sense of
liking something or valuing it because of its beauty. And if tree
swallows choose white feathers over black, then it is similarly
because they value white more than another color. But what if
white feathers are usually more easily seen and found, and the
swallows take feathers of any sort? Since it would be nearly impossible to know what the far-ranging swallows encounter in
the field, it is hard to ascertain whether they indeed have a taste
preference for particular feathers. However, observations in the
wild are generally a sound first step toward deductions related
to natural behavior. It is the place to start. But controlled conditions are then needed to test hypotheses drawn from these observations. There seemed to be no way to know if tree swallows
hunted specifically for white feathers, frequented areas with
abundant white feathers, or picked up any feathers, regardless

of color, based on which ones contrasted most with the background and hence were most easily seen.

I might have left it at that, but for my discovery of a black-and-white domestic road-killed duck. Its feathers offered a perfect control for testing the swallows' color preference. They were all duck feathers, but what variety — black or white in color, short or long in shape, fluffy or smooth in texture. I plucked and stored enough feathers for tests of several swallows in several seasons.

Daily for a week I spread feathers on the garden plot in front of the nest-box while the female was building her nest and the male was singing from the solar panel nearby. I might as well have spread potato chips or tree leaves; no feathers were picked up. Then, right after the female had deposited her third egg, suddenly they were all taken.

It started at 10:30 a.m., when the couple had been perched side by side on the solar panel. I stood fifteen meters from them and tossed out a downy white feather, which drifted to the ground in front of me. No response. But the feathers I'd seen in tree swallow nests before had been mostly long and curved. So I next tried a fifteen-centimeter white tail covert. It had barely left my fingers when the male launched from his perch and caught it before it hit the ground. He flew off with it, but to my surprise he did not take it into the nest. Instead, he flew straight up into the sky with it. There he dropped it, and as it drifted he chased it, caught it, flew on with it, and then dropped it again. Could swallows that live by snagging swift and adroit flying insects be so consistently clumsy at catching and holding on to a slow-moving feather? The feather finally drifted to the ground, where the female then swooped at it once but did not pick it up. There was no doubt, though — they had suddenly both shown interest in a feather, although it looked as if they had used it as a toy.

The pair stayed nearby all morning, and I continued the feather test, using only white feathers ten to twelve centimeters in length. All were accepted. He took eighteen, she only three. All were now taken into the nest. After a while, instead of just tossing the feathers, I held them up in my hand, and the male swooped and snatched them directly from my fingers.

I offered feathers once more in the afternoon, when he took five and she took none. The nest then held three eggs, which rested on a solid base of long, white, curling feathers, the tops arching over the eggs and hiding them from view from above. The swallows had not merely dropped the feathers into the bottom of the nest-box. They had also arranged them in a precise pattern, with the quills of the long feathers placed under the eggs and the plumes arching over them; the arrangement resembled a parasol. The previous nest-building behaviors had suggested that the birds had a notion of what they were doing. They were not just running through an automatic sequence. So what would happen if the feathers were removed? To find out, when the pair left at 3 p.m., I removed all the feathers from the nest.

At dawn the next day, June 1, the male was as usual circling high and chirping continuously, until he eventually descended and perched in the nest-box entrance briefly, as the female entered it. He then left. He returned at 5:25 a.m., flew by the box, and chirped. She responded by coming up to the entrance and peeking out. He made no move to enter the box and left, and she then immediately dropped back down into the nest. But he soon returned and entered the box to join her, and then I heard them softly chirping inside. I had no idea what business he had in the nest except to stay in contact with her, or possibly assess the need for feathers. At 5:33 a.m. he left again and returned in seven minutes, circling and giving a couple of cheeps. She again peeked out, he then left once more, and she again dropped back down into the nest. With both out of view, I then spread a white blanket onto the same level ground used yesterday, and scattered onto the blanket ten white plus ten black feathers, all of them ten to twelve centimeters in length, and curled. I expected the male swallow to return soon, and when he did, he started picking up feathers and carrying them, one after another, into the nest. In a half-hour he had taken all ten of the black feathers, but only three of the white! Color as such was apparently not essential. And background contrast was also a stimulus.

Meanwhile, she stayed inside the box the whole time, and finally after he left I replaced the white blanket with a black tarpaulin, and again put ten white and ten black feathers on it. This time after the male swallow returned, he chose to take and carry into the box all the ten white feathers but only two of the black, again validating the importance of color contrast. The female then finally left the box, having just laid her fourth egg. The male's strong drive to collect feathers was unexpected. More tests were needed — so I again removed all of the feathers from the nest, to prepare for another round the next day.

The male had now taken the lead in feather gathering: what I had seen before was no fluke. A male bird lining the nest made by the female seemed extraordinary, although he may have only delivered the feathers, and she was the one that arranged them within the nest. Also, what had before seemed a definitive choice of white feathers was perhaps not color choice at all. Was color incidental?

Most birds leave the nest after laying an egg. But these swallows repeatedly returned to the nest area and the nest itself throughout the day, during all the five days of egg laying. At least one of the pair was present nearby all the time. The female was often, though not continuously, parked in the nest entrance as though guarding it.

That evening, under a marbled sky and with the temperature at 60 degrees F, she was inside the box at 7:40 p.m. The male circled and stayed almost silent. It took my utmost concentration to try to keep him in sight as he became the tiniest dot in the distance, sometimes coming closer, other times receding. He fluttered, glided, twisted, turned, and circled, until after thirty-one minutes I finally lost sight of him in the dark.

On June 2, the next morning, I got up early. I hoped to arrive before the birds, because I did not want to disturb them in their behaviors at the nest. At 5:10 a.m. the male was circling high above the clearing, and the female entered the box. He continued circling until 5:50 a.m., when he landed on the solar panel after a long swoop down. Then he began to preen, scratch his head by reaching a foot up between a wing and his body, and shake his head. He then scanned his surroundings by turning his head in quick jerks, three to four per second. He took one short break to chase off another swallow. Near 6 a.m. the sun rose over low clouds, and the female left the nest-box and flew away. The nest now had four eggs and also two large white feathers, which I removed to start an extended test. I wanted to see whether the birds might stop gathering feathers after a given time, or after a given number or amount of feathers had accumulated in the nest. I offered some, and the male took only five more, and both he and the female entered the nest-box that morning, when she laid her fifth and last egg.

After that she stayed in the box except once, to chase and then grapple with another female. She had not taken any of the feathers offered (except for the three on the first day of feather lining). It seemed that feather-gathering time had ended, so I stopped providing them. Might the birds still want some enough to search for them?

Twice more before he left his favorite perch on the panel to fly off into the distance, the male swallow first flew to the nest-box to peek in at the female and make soft vocalizations. I felt that at this point his work in the nest-lining process would surely be finished, now that incubation had begun. But on the next day (June 3), while I was on my chair near the nest-box and the male was perched on the panel, he suddenly left it and flew directly *at* me. It was not an attack swoop (a swift descent to the back of the head, accompanied by a rapid-fire series of bill snaps). No, instead he paused to flutter around me before flying on. I smiled, knowing what that meant! He wanted more feathers and considered me a possible source. I heeded his prompt; I got up, retrieved my bag of duck feathers from the cabin, and held a white feather out to him. He instantly swooped down to me and took it from my hand.

The female was at that time in the air, and he flew with the feather to the nest-box, perched in the entrance, leaned in, and dropped it. She then flew down and entered the box, as he returned to perch on the panel and give his sweet chortling song. She answered with similar sweet talk from inside the box.

• • •

THE SWALLOWS STILL had an appetite for feathers by June 4, despite their now full clutch of five eggs. They had again lined their nest, but now the feather lining consisted only of feathers they had found on their own, after I had again removed the feathers from the nest. I did not know where these feathers had come from. The male, however, knew a possible source: while I was sitting outside he suddenly fluttered around my head, and then specifically by my hand. This time I happened to hold not a feather but a spoon (while eating granola out of a bowl). I knew what he asked for, got a feather for him, and, noting his eagerness, brought more.

At one point when I gave him a feather, the female was peeking out of the nest-box, and seeing him with a feather, she flew out after him: she wanted it. A vigorous chase ensued, and he dropped it. She plucked it out of the air and took it into the nest. He then resumed his perch on the panel, and I "fed" him another feather. I was surprised to see him still eager for feathers. I wondered, did he just want to please her?

That evening I watched the swallows exchange places in the nest-box several times; finally, at 7:15 p.m., the male left the clearing for the night and the female stayed inside the box. Just before he left, he circled low over the box and chittered. It looked like a signal that he was leaving for the night.

ON JUNE 5, when I went out at dawn to sit and eat breakfast, I brought along a brown paper bag of feathers. As soon as I sat down, the male swallow flew off the panel and fluttered around me. He could not have seen the feathers in the bag. But his hovering said, as plain as day, what he wanted. And he got it: I again satisfied his feather greed or need, giving him as many as he would take. I thought I had trained the swallow, but instead he had trained me.

I still did not know if the swallows had an instinct to collect feathers during a specific time of the nesting cycle, or if they collected them until they reached a certain number, until the female let the male know when to stop bringing them, or until the male himself had, based on examining the contents of the nest, assessed that there was no more need for them. It appeared to me that the nest was now "full" of feathers — I did not count them — and that the male had indeed stopped collecting. Therefore, I once more removed all the feathers from the nest to continue the testing; the removal of the feathers did stimulate more collecting. Just to be sure, in the afternoon, after again no more feathers were accepted, I once more emptied the nest of all the feathers, and the male again took not only the dozen I had given him, but also brought in three others collected elsewhere: a crow feather, the tip of a Canada goose feather, and a dove feather. When I removed these the next day (June 6), he still took several more from me to bring into the nest.

Three days later, in the morning of June 9, the now-incubating swallows had in their nest not only the three white duck feathers I had given the male on June 6, but also eight new ones that they had found on their own. Four of them were likely from a wood duck, and the others probably came from a Canada goose, and one from a great blue heron. A beaver bog three kilometers to the south, where I had seen the swallows fly routinely and where several pairs of herons nested, was the likely source.

Incubation requires that one parent forgo foraging time, and so for most birds, when both must later stay around to feed the young, the pair takes turns incubating. Among some types of birds, one bird does all the incubation and the other forages and feeds the mate. I had not seen a tree swallow feed its mate, either as a nuptial offering before nesting (which commonly serves as a test of a potential mate's reliability as a provider and hence as a suitable partner) or during the egg-laying and incubation periods. Did the male and female tree swallows take turns incubating? Now that my pair's clutch was complete, I'd find out how they balanced meeting their various needs by determining what each of them did.

A snippet of less than a half-hour on June 9, one of the warmest days so far (78ºF in the morning and up to 87ºF by afternoon), shows the process: At 7:10 a.m., the male vocalized and flew toward the nest entrance, and the female popped out of the nest-box hole as he approached. She flew over the clearing and voided a white fecal pellet. In less than a minute she slipped back in, and he left. In a few minutes he came back and again perched at the nest entrance after she came out, flew high, and circled. He meanwhile stayed at the entrance and peeked out, without going down into the nest. She flew high around the clearing and at 7:30 a.m. dived down. He was then still at the entrance and had been peeking out, but then slipped out and flew off, allowing her to zip in. The timing of one out and the other in was so close, you might have expected a midair collision in front of the nest-box.

The male chirped when he came zooming in, so that the female could pop up to and out of the entrance a millisecond before he reached it and shot in. But sometimes she stayed down when he came, and he then entered the nest-box, came back up to the entrance, and peeked out while she stayed down on the eggs. Thus, he spent time at the nest-box, but when he entered it, he almost immediately perched in the entrance to peek out. Her incubation breaks were frequent and usually lasted only two to four minutes. The two were taking turns in the nest-box but not in the incubating.

On her seventh day of incubation, the nest contained only five feathers. It seemed unlikely that the swallows would still have interest in procuring nesting material, but I tested them never-theless to make sure, finding that the male no longer accepted from my hand the favorite long, white, curled-over feathers. Nonetheless, I put ten of them on the ground, and the next day seven of them were in the nest. The swallows ignored the five new feathers offered the next day, and I then removed all from the nest for one last test.

On June 12, a cool dawn (49ºF) followed a night of constant drizzling rain, and I placed ten of the twelve feathers I had re-moved from the nest on the ground in front of the male, perched, as usual, on the solar panel. He had collected feathers on the warmest days, about 85 degrees F, when there was no need for insulation but instead a potential problem of overheating in the clay nest-box, which was in direct sunshine. Furthermore, there had been no energy crunch because insects, the bird's fuel for heating and exercise, were now abundant. But today it was cool. If those large curved feathers help retain heat in the swallow nest, then the male "should" pick them up off the ground where I had placed them and carry them back into the nest, to serve as a baby blanket. But by 7:30 a.m. he had not made a move to do so. There was still a light rain, and the temperature still held at 49 degrees F. He had before been the primary provider of feath-ers, so if there was any call for feathers, he should have answered it. Was the time for nest feathering finally past?

The cold continued the next day (June 13), and it was also wet and windy. The spring peepers were still piping shrilly in the vernal pool. A yellow-throated warbler sang, and the phoebe gave its usual early morning reveille. In the near dark of the dawn, the male ruby-throated hummingbird was at our feeder, fueling up on sucrose. At 8:30 a.m., when the female swallow left the nest-box for a short break from incubation, I checked its contents, and to my surprise saw two new feathers added since yesterday afternoon. I removed both and then, while she was still flying around, held out the lighter-colored one to her. She came by me; *she* had interest. I put the feather on the ground, and she picked it up and returned it to the nest. Her interest induced me to then lay out ten of the large white feathers. The male ignored them, but the female took four of them, and the next day she took three more. He showed no interest in them at any time.

Later, when he was perched in the nest entrance, I walked up to him and held a long white contour feather directly in front of him. He did not fly out to get it, so I edged closer, stood still, reached over, and touched his bill with it. He shook his head in mild irritation and continued as before to turn and look around, as though I did not exist. He stayed in place a couple more minutes after I backed off; then he flew out to the solar panel.

The female's flights to and from the nest-box became unusually frequent over the next two days, and so at 10 a.m. (on the 15th) I looked into the box. A baby had hatched. Naked and pink, except for a few fluffy white down feathers, it squirmed as it lay among the white eggs. The incubation period was now ending, and the female had taken in feathers throughout it.

7

TENDING THE BABIES — 2014

THE PLANNED EXPERIMENT WAS FINISHED AND HAD yielded more than I could have hoped. There seemed little to gain by continuing to watch the swallows this year. But the lure of seeing something new or unexpected is powerful, and this usually requires being at the wrong place at the right time or vice versa, mucking around in one's much-loved places and situations. It seemed the right time and place for watching baby swallows grow up, and now, on an unusually pleasant sunny day, as it warmed to 68 degrees F and insects buzzed in the air, was the time to start. When I opened the nest-box, three newly hatched and still-blind babies raised their heads and briefly opened their tiny yellow-rimmed bills to show their gaping pink mouths. They had felt the box disturbance that signaled "food is coming."

The following day started cool under a clear and windless sky, and I was settled in five meters from the nest, relaxing with a cup of coffee at 6:03 a.m. With nothing to expect beyond the ordinary, I was loaded with patience and watching to see what might happen next as the male perched in silence on the panel and the female remained in the nest. But after a few minutes he flew over to hang at the nest-box entrance, and she then came up from inside the nest and flew off as he slipped in. A few seconds later he popped back up and perched in the entrance. A minute later she came back and he flew out a millisecond before she zipped in to settle down into the nest again, as he returned to his panel perch. It looked like a continuation of the old routine, now that all but one of the eggs had hatched yesterday, but I wanted to keep tabs and make sure.

The male left the panel after only a minute and didn't return until 6:44 a.m., when he entered the box and the female left it. She flew at a frenetic pace all around the clearing but quickly returned into the box, then popped up to the entrance and perched there to look around for another minute before leaving again as he returned. This pattern changed little. Keeping the babies warm and incubating the one remaining egg were the priorities.

The coffee cooled, and time seemed to stand still against a backdrop of bird voices in all directions, a reminder of life pulsing along while I was busily taking notes and the swallows were whispering to each other. In the background, a hairy woodpecker endlessly repeated his monologue of second-long drumming rolls, one after another, until I hardly heard them at all. The phoebes made an occasional *chip* as they flew from one pole to another on the garden fence, and then fluttered up to their nest under the roof over the window, to feed their young. As they had yesterday, one of the pair of phoebes flew *at* the other in what looked like a nudge. The intent can only be guessed at, but the main point of interest was that some intent was apparent. A scarlet tanager sang several brief notes. A ruffed grouse drummed once. Red-eyed vireos sang without a pause from three directions, and a black-capped chickadee called near a dead poplar tree, where he and his mate had excavated their nest hole. The fifth and last of the tree swallow eggs hatched in that hour, more than a day after all the others.

ON JUNE 18 THE BABIES were snug in their nest of now mostly, but not exclusively, dry grass. The parents had for the past two days ignored the feathers I had offered, and they now hunted almost constantly for insects, landing only to feed their young; one of the parents stayed at the nest while the other hunted.

Surely the feather gathering would be long past. But just to be sure, I removed the remaining ten feathers from the nest and placed them on the ground, and the female scooped one of them up but then dropped it — but later picked up two and brought both into the box. This behavior seemed extraordinary, now that the main effort must be to hunt insects to feed the young. But the swallows had time to spare since the weather was fine.

The next dawn I sat down to watch several exchanges at the box, as the female shot in and the male out. She stayed inside the nest-box for nineteen, thirteen, six, and nine minutes, as he waited while perched in the nest-box entrance for one, two, and two minutes.

The exchanges went like clockwork. When she was in, she stayed down. After he slipped in, he almost immediately came back up to perch in the nest-box entrance. He watched her from there as she circled around the clearing, so when she dived down to get back into the box, he was already out a fraction of a second ahead, and she could continue her dive into the box.

All of this was by now predictable and familiar, but I was still writing it down because one never knows what might come up next. Hours passed quickly as I kept my eye on the swallow nest-box where the male had perched motionless since 6:52 a.m. Then in a flash everything changed: a swallow I at first presumed was his mate came flying in toward the nest-box, but this time he did not budge, as he had always done, to let his mate in. This nonresponse was, in the context of my hundreds of prior similar observations, unprecedented and therefore of significance. *This* swallow, after passing the nest-box, made a low circle, then zoomed by it, and the male blocked the entrance once more. And *again*, he did not budge to make way. This newly arrived swallow could not be his mate. A whirl of questions: Who was this? Then *two* swallows flew overhead. What did they want? Agitation calls sounded within seconds. The male remained perched in the nest entrance, as he always did during nest guarding. Then both of these new swallows flew down to him at the box entrance. But he still stayed put, even as they both fluttered directly in front of him. He showed no visible reaction. Seconds later, two females engaged in a violent chase.

This "strange" female could not have missed seeing the male looking out the box, and he showed that he knew she was not his mate. Yet for a potential nest cuckolding, it was far too late in the season. The male swallow had nothing to defend from that quarter. However, perhaps thinking of tree swallow behavior in terms of only the present could be misleading; there is always the beckoning possibility of the next year.

Like most birds, tree swallows try to return to their previous nest sites. (Their maximum known life span is twelve years.) But often they don't find a free nesting place when they come, and up to half per year don't make it back. Perhaps for a swallow (or a pair?) this site was now lost in this year's contest, but it might become available in the future. Exploration now could smooth the path to a destination for the coming spring. The resident male has nothing to lose by tolerance now, but he does have much to gain if his female fails to return. And for birds living on average fewer than five years, that is always a strong possibility.

The second pair came again at 6 a.m. (June 20), when for minutes they flew lazily around the clearing while the nesting pair perched nonchalantly on the panel. Their babies now had the "six o'clock shadow" of dark pinfeathers. They were growing fast and nearing adult weight.

Much eating is associated with much excreting. The babies' excreta can, in this type of cavity nest, be disposed of only if the parents can enter the nest-box to reach them. By late afternoon I had tabulated that the male had done nine shit carries, the female none. He was always in and out in less than thirty seconds, while she stayed at the entrance for three, six, and up to ten minutes at a time. Such taking of leisure at a time when rapid provisioning of food was crucial indicated that the local crop of insects was now more than adequate.

On the next day (June 21), the longest day of the year, the weather was still ideal for flying insects. The pair was almost silent, except for some brief chirping from within the nest. The female tended to loiter at the entrance. The male continued to carry out the waste. A third swallow came, and it flew around with the pair as if they were friends.

No baby monopolized the nest entrance to intercept most of the incoming food, as had happened before at other nests, causing the babies to starve. These young were all getting fed, grow-

ing at the same rate, and seemed set to fledge nearly synchronously.

On another gorgeous morning (June 23) the babies now had
feather stubble sticking out all over. In one inspection of the nest
I found a mayfly jutting out of the side of one nestling's mouth.
Only one leg of its normal six remained, and its three long tail
bristles were missing, but enough was left for me to tentatively
identify it as *Ephemera simulans*, the brown drake, well known
as bait for trout fishing. The swallows had likely been making
foraging trips to the nearby beaver bogs, streams, or lakes, where
earlier they had collected feathers.

The female made nine foraging trips to the male's seven, and
in another hour she brought off seven trips and he five. However, a second female again visited the nest-box that morning.
Normally the resident female literally dived in, but this new one,
as before, only fluttered at the entrance and then veered off and
came back again. She landed at the entrance, hung there, and
only then entered the nest-box. She stayed inside for only twenty
seconds before she shot back out.

The next morning (June 24) the new female came again (at
about 11 a.m.), but this time it was different — the pair happened
to be inside the nest-box. Twice she flew by the entrance, but on
the third approach she landed on it and looked in for about fifteen seconds before leaving. It was only then that the male left
the box. Not quite believing my senses, I opened the box and saw
his mate still inside. Neither of the pair had made any response
that I could discern to the visiting female. With both females
close together, it had been easy to compare them: the resident
female had a brown head with faint blue on the temples, but the
visitor's head was all brown. Who was she?

This new female was not an offspring of the pair from the previous year (2013); no young had fledged then. Was she perhaps the swallow that had been visiting the nest throughout this nesting season? Did she have some vested interest here? Had she left an egg in the nest? Was one of the five young hers? Was she interested in the male? Whatever the reason, the resident pair tolerated her now. And why not, if she could be a potential partner to the male in the future, or a potential helper now, bringing in food for the young? On the following day (June 25) during my hour-long morning watch, she again visited and went briefly into the nest-box.

The resident male made seven trips per hour, his mate ten. The nest was then lined with eight feathers, and food kept flooding in. The young were still the same size, weighing about twenty-four grams each, or four grams *more* than an adult (as per my bird guide). Yet though they were well fed, it was unlikely that these young could be considered "overweight": they would need a cushion of fat to tide them over while beginning to forage on their own, almost from the first day out of the nest.

By June 28 the five young were fully grown and garbed in ash-gray feathers. Their behavior showed that they had been well fed. In previous years, the babies jumped to the nest-box entrance and then stayed perched there, stridently begging until a parent brought food. This year the young stayed down in the nest and remained silent, except for making a brief response when a parent arrived. The female now made six, and the male eight, food deliveries per hour. But the pair sometimes tarried, gliding around the clearing after feeding the young, taking time to perch on the panel, and sometimes parking in the nest entrance after a food delivery. They were easily keeping up with the job of providing nourishment and did not feel under stress to accomplish it.

On the morning of July 1, the male came only four times in an hour, the female only three. The parents' diminishing feeding visits to the nest, despite the continuing good weather, meant the babies were about to fledge, and I was anxious to see the process. The other female now came eight times during that same hour, but she didn't enter the nest-box. The resident pair then became downright lackadaisical; the male perched for long durations at the nest-box entrance or on the panel.

The young were now starting to beg, and they cried louder when a parent perched at the entrance. But they remained silent in the presence of the other female; her call when she approached the box was more "ringing" than the parents' call. She even landed at the entrance once while the other female was still inside; the resident female came up to the nest hole, but instead of chasing the visitor off, she merely ducked back down into the nest, as the other female perched at the entrance. A pair of chickadees also came by. One landed on the top of the box and called, then went to the nest entrance and peeked in. Its mate came and did the same. They too were ignored. It was two months past the time when both chickadees and tree swallows begin a nest, so neither bird posed a threat.

The chickadees were no more helpful at this nest than the other female swallow appeared to be. Both were perhaps faintly expressing instincts that would rage in intensity at the appropriate time — perhaps like the swallows' instinct to gather feathers to line the nest, which also was expressed at a low ebb even after nest-building time, when it was most useful. In contrast, the parents' diminishing nest visits to feed the young appeared strategic. It concerned the immediate — working up the young swallows' motivation to leave the nest. Why did they not let the young leave on their own, when they pleased? It must be because the adults needed to lead or protect them.

By afternoon (from 4:39 to 5:39 p.m.) the male and the female both made only two nest visits, although they perched at the entrance for eleven and nine minutes, respectively. Curiously, today for the first time they dived at my head, something they now did when any animal approached. They were becoming defensive in preparation for the imminent exit of the young; the chicks would be vulnerable to predation by hawks. When in the late afternoon I opened the box to check on the young, two of them scuttled up to the edge of the opening as if wanting to fly out. I blocked them and closed the box. No adults were near, and the babies stayed inside.

The next morning (July 2), it was warm, wind-still, and clear, a perfect day to sit and watch the swallows fledge. At 6 a.m. the parents were perched on the panel, and now one of the babies was peeking out of the nest-box. All were quiet. There was nothing untoward or unusual to raise my expectations. But just such a moment is when the unusual can happen. I had watched the family this long. I would watch them to the end.

FIFTEEN MINUTES LATER, first one and then the other of the pair of swallows left, then returned in several minutes to again perch on the solar panel. Eight minutes later the female flew several times around the box and again returned to the panel. A baby was perching in the box entrance. It chirped and started to jerk its head around as the female flew by, but it continued to park there, even until 6:49 a.m., when another pair of swallows arrived and joined the parents.

Three swallows were soon flying around, while one remained perched on the panel, but then the three landed, and all four perched within a meter of one another. I could not distinguish the newcomers from the parents, except that one of the females had a bulging, food-stuffed gullet. Insect legs poked out from the corner of her bill. She eventually flew to the nest entrance and perched there in front of a baby, but she did not hand the food over. Instead, right there on the spot, she ingested it herself. Then the male also flew to the nest entrance. He made a little peck toward the baby's open bill, as though faking a food delivery.

Two adult females next to one male perched close together on the panel in apparent harmony, and one of them made a tiny nudge against the other, who jumped up to perch a meter farther away. But it seemed like a gesture, not aggression. The nonresident pair left at 6:57 a.m., and the male then immediately gave a brief rendition of his sweet chortling song, which I hadn't heard since the eggs were laid. The pair also made soft whispering calls to each other as they perched side by side on the panel, preening, shaking, stretching their wings.

Twenty minutes later, one of the babies finally started making loud chirps while leaning far out of the nest-box entrance. An animated conversation of chirps then ensued between the chick and the pair on the panel. All of a sudden the baby, apparently quite spontaneously, jumped out and started fluttering toward the woods. Both parents instantly left their perch and accompanied it, making passes over and around the fledgling until it landed in the heavily leafed branches of a maple tree. A second baby then came up to the nest entrance, and both parents made numerous flybys. Much vocalization ensued. The parents brought it no food, though, and ten minutes later this baby also leaned out farther and seemed more animated; then it fluttered off. The pair accompanied this one also, until it landed in another thickly leafed tree at the edge of the clearing.

Then there was total silence at the nest-box. No new baby appeared at the entrance. I was baffled. The two exits had been conspicuous events — how could I have missed the exits of the other three babies? It seemed improbable that they had left; I had watched continuously since dawn. Nor did it seem probable that the last three would now be silent, after the long period of forced fasting. The parents nevertheless flew to the now-silent nest-box. There was no response from within. Weird. The female hung at the entrance a few moments, then flew back to each of the separate trees in which the two fledglings had landed. But not to any other trees. She circled the clearing, returned to the nest entrance once more, then circled again, and cheeped loudly. She circled high, came back down, and again peeked into the box. Back and forth she flew between the box and the two trees, always noisily calling. The second pair had returned, and all four adults, as before, again perched for long durations near one another on the panel.

I had not approached the nest-box during this time, to avoid interfering with the fledging. But with the continuing apparent agitation of the parents and the deafening silence from the nest, it seemed that I had no option but to check it, and as soon as I opened it, I experienced shock and disbelief: three ready-to-fledge babies lay dead in the nest.

Their belly feathers were caked with fresh red blood, still

partly liquid. Their undersides looked macerated by ragged tears, as if they had been chewed. As I lifted one of the nestlings, a large black beetle, with needle-sharp pincers, dropped into my hands, and it tried to bite my finger when I grabbed it. It was a *Nicrophorus* burying beetle, an almost all-black species with only tiny blotches of orange at the end of each wing-cover. For decades here I had seen hundreds of burying beetles because I had studied their behaviors by setting out animal carcasses for them to feed on and then bury as food for their larvae. But I had never before seen a single specimen of this particular species. Its distinctive coloration positively identified it as *N. pustulatus*. Burying beetles, or sexton beetles, are named for the way they bury dead mice, shrews, and small birds. But *N. pustulatus* has the singular habit of burying turtle eggs instead, using them, rather than animal carcasses, as food for its larvae. Finding *this* beetle, in this context, was beyond bizarre.

An autopsy revealed nothing to solve the mystery. All three baby swallows had conspicuous pads of body fat, so they had not starved. Each one's stomach contained undigested insect remains. They weighed thirty grams each (compared to the published weight of twenty grams per adult). These young had been in prime condition. As the fresh blood showed, they had been killed within the past few hours, that is, the previous night. These babies looked mauled. Had a weasel or a squirrel entered the nest-box? If so, why did it leave its prey behind? No larger predator could have entered the box. I saved the rare beetle, the dead young, and the nest contents so I could search for potential clues to the baby swallows' deaths.

Experts informed me that the beetle would not and could not

have been the killer. No carrion beetle would attack a live bird. The bird would have to be dead and starting to decay before a beetle would be attracted. The fly larvae in the nest were identified and excluded as potential culprits; some feed on fecal material, and others may take blood, but they don't make one-night raids for it. Nobody had seen anything like this — I was in effect presumed to be hallucinating. But I had peeked into the nest-box the previous evening and had seen the young looking healthy. If the rare beetle offered no explanation, what could? There were no apparent clues, but on further reflection I recalled another piece of evidence I had at first not deemed relevant: the disruption of the nest itself.

Could a raccoon have reached in and "fished around" blindly with a paw? Raccoons have sensitive paws and regularly forage, by tactile sense, with them — feeling around in water under logs and rocks in search of frogs, crayfish, and other prey. One could have climbed the wooden pole and extended a forelimb into the nest-box entrance, to reach partway in. The young swallows, close to fledging, would have fluttered about in panic. They had no way to flee. Exhausted from being batted about, they might have died on the bottom of the nest-box. The beetle could then have been attracted by the scent of their blood.

Because turtle eggs are normally buried in soil, turtle-egg-specialist beetles locate them by scent, not sight. The buried eggs are alive, so these beetles are not attracted to the scent of carrion. This idea of scent offered a clue, though. Most eastern turtles live in bogs, and tree swallows feed prominently on insect hatches from bogs. The scent that tags the turtle eggs buried in sand perhaps also tagged the baby swallows in the box, thus attracting the egg-seeking beetle. It had found itself at a strange feast indeed — the equivalent of baby birds confusing red candy gumdrops for berries, and actually eating them.

IN THE MEANTIME, the adult swallows had, by the next morning (July 3), not forgotten their long-silent unfledged babies. It seemed amazing, given their prior frequent entries into the nest-box containing the carcasses. It had poured all that night, and a thunderstorm had blown in. A thick fog enveloped the clearing, but the pair came just the same. For ninety minutes, from 6:30 to 8 a.m., each of them called and circled ceaselessly. They sometimes landed in the heavy wet foliage of trees, which they had not done before. At times they also perched on the panel, but they no longer revisited the nest-box (I had removed its contents the previous evening). The second pair of swallows also arrived (at 8:20 a.m.), and as before this pair again landed on the panel, next to the parents of the three unfledged babies. By 10:30 a.m., however, all the swallows had finally left.

While the parents had come looking for their young, the second pair of swallows had taken a seemingly extraordinary interest in the proceedings. It was understandable that the resident pair did not now behave aggressively toward these other swallows. The extra pair's apparent interest throughout the nesting cycle strongly suggested that they had an unfulfilled nesting instinct, which they exercised vicariously. They were likely among the many swallows that had not secured a nest site this year, but they may have cultivated a potential one for the next.

8

A NESTLING REDUCTION — 2015

ALMOST INCONCEIVABLE HARDSHIPS ARE FOISTED ON some animals, including the human kind, through excess reproduction. The effects become apparent only in retrospect, when it is too late to correct. Some birds, however, have evolved behaviors, likely unconscious, that anticipate and help ward off or reduce problems related to numerous offspring. It starts with the number of eggs, which is automatically held to a rough standard per individual.

Birds need to feed their fast-growing nestlings huge amounts of food per day, and the number of eggs laid is often determined by the amount of food available to the female early in the reproductive cycle. Snowy owls may lay clutches of over a dozen eggs in years when lemmings, their main source of food, are at the peak of their population cycles. In lean years the owls lay only several or even none. Brood reduction can be achieved even after the eggs are laid, or even after the babies are hatched. For some eagles, only one youngster can usually be raised per clutch, but the usual clutch consists of two eggs, one serving as insurance in case the other does not hatch. If both hatch, then one of the two often becomes a meal for the other. Among some herons, one sibling attacks and ejects a weaker (because less well-fed) sibling from the nest. Cuckoo nestlings, hatched in the nest of some other species, routinely roll their host's eggs or shove the other nestlings out of the nest. The tree swallows nesting in early spring, when it can be cold and rainy, with few flying insects available as food, routinely face periods of starvation. It seems logical that they too would have evolved ways of reducing the size of their own brood when starvation threatens to kill them all.

Overproduction is particularly costly to females. Margaret McVey, a researcher of barn swallows on the Isles of Shoals, off the coast of Portsmouth, New Hampshire, followed the returns of individual birds to the nest each season. She found that about 50 percent fewer female swallows returned after migration than males, likely because of their much larger investment in egg rather than sperm production. Some females "disappeared" after a week of rain, and McVey found one female that died directly on her nest, with nestlings. The bird's mate continued to try to feed the young, but they starved because he could not feed all of them. Once, after it had rained for five days, McVey found a fresh barn-swallow egg on a nearby lawn. A female had limited her clutch by leaving it there. If she had tried to feed the full clutch of about a half-dozen young, all could or would have starved. Similarly, I once found, after a week of heavy rains, the number of eggs in a blue-headed vireo nest I had been monitoring decreased by two eggs.

The ability to reduce a clutch number confers a large reproductive advantage and would therefore be expected to evolve, and it seems less draconian to sacrifice eggs, rather than live young. But the latter option is routine among some herons and eagles; one or more young are pushed out of the nest or eaten when food is scarce. Sometimes humans feel pressure to make a similar sacrifice. I won't forget one incident in which a father, in the winds of war, gave this advice to the mother of his children: "try to save at least one of them." He was trying to be an optimist, and moral.

COLD WINDS BLEW during early spring 2015, and April 15 brought the first warming after three days of rain and snow. A thick crust of ice on deep snow still covered the clearing. But the night sky cleared, and by 9 a.m. temperatures had risen to 55 degrees F. Swallows may have used the southerly winds to come back. Suddenly there it was — a male in shiny blue-green nuptial plumage at the entrance of the same nest-box used last year. He peeked into it and then entered and stayed inside over two minutes, then flew from one box to another. His behavior suggested that he had come back home. I knew that observing more details of his activities should give more clues.

Over the next three days he came always alone and visited the same nest-box. He ignored me when I approached him as he perched alone and silent on the solar panel. No female arrived until nearly noon, four days later. She looked more glossy than last year's female, either because of a molt or because she was a different individual, with a more greenish cast. Her chest, slightly smoke-tinged in color, contrasted to his, which was gleaming white. They flew together and also perched next to each other on the panel, but not in close proximity. He did not vocalize when flying (or possibly mildly chasing her?), but when they returned he sang softly and then flew to and kept entering the same nest-box. She eventually flew down and perched at that nest entrance, but she acted skittish and only peeked in.

The male's reaction to this female was somewhat lukewarm; he seemed to tolerate her while perhaps waiting for his mate. However, on the next day, April 17, the two perched on the locust tree, now only about three meters apart rather than at opposite ends of the solar panel as before, and although both sailed around independently, they returned to the same tree. Later she landed at the nest-box entrance and hung there for several minutes, while he chirped his apparent satisfaction and approval from the panel. As soon as she left the box, he stopped chirping.

He was alone on the panel on April 18, from at least 5:20 to 7:10 a.m., when a second male swallow arrived. A vigorous, prolonged, but silent chase ensued, and the males then tangled violently in the air, becoming enmeshed, fluttering to the ground, and then leaving. Then the female that had been present yesterday (or a different one) also arrived, and the male then flew to the nest-box entrance, not to guard it but to guide her to it by his singing. Again he only ducked his head in, as if making a gesture, then returned to his perch. She then flew to him, and they perched a half-meter apart.

By 9 a.m. she entered the nest-box he had shown her, and she stayed inside for about two minutes before coming back out. The pair then briefly perched ten centimeters from each other on the panel before she left, landed on the ground in the garden, and started yanking on stems of dry grass. She loosened a piece from the soil and flew around the clearing with it, then finally entered the box with it in her bill. She seemed an expert at this behavior right from the start, in stark contrast to the behavior of the female the year before. It seemed that this glossier female, clearly not last year's, had accepted the male as her mate. Nesting could have started then, but weather intervened, and two rainy swallowless days followed.

DESPITE THE EARLY MORNING frost on April 22, a swallow flew high at 6 a.m., a barely visible dot moving in huge erratic circles around the clearing, chirping continuously. I lost sight of it after five minutes yet could still hear its nonstop song. At 6:17 a.m. it descended — I could now see it was the male — and landed on the locust tree, next to the female. He stayed by her until 6:35 a.m. and then flew down to perch on the solar panel, singing there in repetitive, soft, three-note liquid chortling phrases. She remained briefly on the same spot on the tree but then joined him on the panel, and he responded with several more of his two-to-three-note soliciting calls before flying to the nest-box, hanging from the entrance, and looking back up at her. Then he slipped inside. She then also came to the entrance, and he returned to the panel, where he chirped again but with more vigor, as she continued to hang at the nest-box a few minutes more; she finally entered. This time she stayed down for eight minutes. After she came out, she flew out of sight, and he stayed on the panel a couple of minutes more before leaving also. In five minutes she was back on the tree, and he joined her. I was now convinced that they had become a couple and were united on where to nest.

But then the next day (April 23) it poured, and at night the rain changed to snow. At dawn the ground was covered, and snow continued to come down. The pair stayed on the locust tree until 10 a.m. and then left together, not returning until three days later, on April 26, a beautifully still and sunny morning, with temperatures rising to 46 degrees F. On this and the next two days they arrived together at about 6 a.m. every morning, to first perch in their locust tree for about a half-hour. The male then sang from the panel. She entered the nest-box, and he visited it briefly, with her inside. They would leave, then later both revisit the nest-box and stay in it for minutes at a time.

Two days later, April 28, things got more serious: the male fluttered five times over the female's back, in mock mating. She did not comply by assuming the crouching posture. The mock mating was followed by repeated nest-box visits and perches on the panel or the locust tree. Finally, near 8 a.m., she flew to the ground in the garden and picked up a piece of grass, brought it into the nest-box, and stayed inside for five minutes as he sang; then they left. They had sealed the deal: they would nest, and I could now look forward to seeing the process unfold.

The next day, the first day that a local pond was free of ice, the pair came despite an overcast sky and the low temperature of 40 degrees F. A third swallow arrived as well, and after a quick violent chase of the intruder, the pair returned to their perch on the locust tree.

As per the usual pattern, both swallows arrived the next morning by 6:30, and when a third appeared, the pair immediately chased it off. The two then perched on the locust, and the female started swooping low over the ground in the garden before landing there several times. On one foray she picked up a long piece of grass and carried it into the nest-box. But the main event this morning occurred when the male, while perched alone on the locust, suddenly took off, making excited chase calls as he disappeared almost out of my sight, pursuing a distant swallow. Then, later, another swallow came flying over and was immediately accepted. Did the male recognize individuals even when they were so far away, I could barely see them? Keen discrimination of individuals would not evolve unless there was strong advantage to it. The next dawn (May 1), the pair came early, and as usual a third arrived and was chased off. The female again picked up nest material in the garden. At dawn the next day there was frost on the ground, and the male was already circling high, repetitively giving a double-note call. After a while I saw the female too. The early sun glinted off their white bellies. He descended and landed on his usual twig of the locust as she kept on circling. He flew back up to join her, then came down to the same perch, as she followed and landed beside him. Three times in rapid succession, he then hovered over her in mock mating — I saw no contact. And again they rose into the air for more high circling.

This local pair interacted with possibly two different swallows, but I could determine only that they sometimes flew together. In one group flight, three swallows stayed together, passing over the clearing, from about 10 to 10:30 a.m., making excited *chee chee* chatter. One bird eventually settled on the solar panel, and another immediately perched close beside it. This, I suspected, was the resident pair, and only now did the third one leave; it was not chased. Several minutes later the pair flew up, and then four swallows were flying together. Again, one dived down and perched a moment in the nest-box entrance, until one of the remaining three chased it vigorously and the two tangled in the air and tumbled down to the ground in a brawl. All left afterward, but the local pair returned in forty minutes. The male again perched on the usual spot on the black locust tree.

New grass was sprouting by May 3, and at 8 a.m. the female was carrying last year's dry grass into the nest-box. The temperature rose to 82 degrees F the next day. Black-cherry leaf buds were opening, and red maples and coltsfoot were blooming in glorious red and yellow profusion. Spring was happening, and the female was busily gathering grass, quickly forming a crescent of it on the nest-box floor. For the next seven days the pair of swallows continued to arrive before sunrise, while green was bursting out all over, transforming the woods. Hummingbirds were back, feeding on sap licks made by sapsuckers.

But now the ring of grass in the swallow nest-box was no longer growing in size; progress had for some reason slowed or stopped. A third swallow arrived at times and was tolerated, and the three even perched near one another, but chases too had continued. Had the third swallow somehow interfered with the nest building?

By May 13, however, the nest looked nearly finished, except for the feather lining. Feathers had not been in the forefront of my thoughts, given the many unexpected observations that spring concerning the swallows. But now, with the nest all but completed, I could take another look at this feature of nest building. Did the swallows really prefer white feathers, or were they simply the easiest to find because they showed up better against most backgrounds? My previous tests may have overemphasized the effect of contrast, without giving choice enough consideration. I still had the bag of black and white duck feathers. This time I would alter the test by simultaneously laying out a white tarpaulin and a black tarpaulin side by side and give the swallows time to get used to them before starting the experiment.

On May 16 I placed a single white and a single black feather on the white tarp, and the same pairing on the black tarp. Given simultaneous choice of background and feathers, which feathers would prove popular? The swallows (both male and female) were ready and eager. The results at the end of the day: twenty-six white feathers were taken from the black tarp, four white from the white tarp, fourteen black from the white tarp, and three black from the black tarp. As in the past year, background made a difference, but overall, with the background effect evened out, more white feathers were taken than black. In a repeat of the experiment, on May 19, an overcast and drizzly day, both birds were less interested; overall, only five white and nine black feathers were taken, making a total of thirty-five white versus twenty-six black. It is not much of a difference, but it implies that contrast wins out over firm preference for a certain color. White feathers are more visible than feathers of other colors against almost any background in a field or forest; the presence of mostly white feathers in a tree swallow nest could have evolved over time for this reason alone, without there being any preference for it.

Finally, at dawn on May 20, while perching on the locust tree, the swallows mated, eight times. The female then picked up a white feather, flew to the nest-box entrance, went in, and stayed down for forty minutes. The first egg was laid the next day. The pair remained long hours perched on the locust tree, and the pair once chased a third bird away. If this was not actual nest guarding against parasite eggs, it was at least a convincing demonstration of what that *should* look like.

On May 22, in the semidarkness of dawn, a swallow was chirping high above, and after searching for it in the overcast sky, I saw it fluttering. Ten minutes later there were two, and the male came down to perch on his spot on the locust tree, while the female flew into the box — to lay the second egg, I presumed. Three minutes later the male swooped over to the box, perched at the entrance, went in, and joined her there, as a faint chirping ensued but quickly subsided. I approached the box and put my ear next to it, still hearing a barely audible high-pitched twitter. I continued to watch, not letting my eye leave the entrance for a second, because I wanted to know how long the male would stay with his mate. It became very still, and then to my surprise a swallow appeared silently out of the sky to land briefly on the locust tree, and then circle. Was this a third swallow, or had I missed seeing one of the pair leave the box?

I had up to this point presumed that the pair were still in the box, since I had not seen a swallow leave. Then at 6:45 a.m. the male did finally leave the box, having been inside for ten minutes. He joined the newcomer — the two flew together without antagonism. They then left together, as though a couple. But by my accounting, the female should still be in the box. Not wanting to disturb anything, I had been merely watching carefully. But now I could check the box. I was flabbergasted: the female was inside! She was almost hidden by all the large feathers. I reached in, pushed them apart with my fingers — and there she was, hunkered down, seemingly at rest. She did not budge, nor did she fly out after I closed the box. Her mate then returned, after a two-minute absence, to his perch on the locust, and she then flew up to him, and they both left. I then checked the nest again: the second egg had been laid.

On May 23, the pair arrived early, as before, and the third egg was laid by 8:05 a.m. They again mated, and both perched together on the panel. Suddenly the male flew to the ground and picked up a long white feather I had left there. But instead of flying into the box with it, he carried it up into the sky. The female left her perch and followed him. As he gained altitude, he dropped the feather, and it drifted in the wind. She caught it, and dropped it also. He made a weak pass toward the feather as it floated to the ground.

The next morning, May 24, the fourth egg was laid. But by evening the male was alone on his perch on the locust, and at 7:50 p.m. he flew from there into the nest-box. There was then whispering-chittering in the box, and I stayed to see whether the couple might leave during the night. But neither had gone by dusk, at 8:40 p.m.

On May 25, I was awakened at 4:55 a.m. by the continuous chirping of a swallow, and I rushed out to the nest, which still contained the four eggs. They felt cold to my fingertips. The birds had perhaps not spent the night in the nest. The continuously vocal high-flying bird was a male (as I learned later), and he continued chirping while circling high. At 5:23 a.m. the female finally arrived, and he then immediately sang his chittering-gurgling song. Now they both circled, but she soon swooped down to zip by the nest-box, and he stopped his circling and sang all the more. Both soon landed on their perch on the locust tree to mate there, after which she flew off. But a third swallow now came, and all three circled together. One left in less than two minutes, and the male resumed perching on the locust and sang, as the female circled and again flew by the nest-box before returning to perch next to her mate. Both then sat in silence. At 6:37 a.m. he started to sing, and she again flew down and past the nest-box several times in succession, before landing on the entrance and hanging there, looking around. She seemed hesitant, but finally slipped in. All was then quiet, but at 6:55 a.m. a third swallow came, and the male left his perch to meet it. They flew around peacefully as before. This third swallow left within a minute, and the resident male then flew to the nest-box entrance and peeked in briefly before resuming his perch on the panel. I then checked the box, and as expected, the female was inside, and the nest contained the fifth egg.

By May 26, incubation was beginning, a new chapter in the swallows' lives. Might it also mark a new stage of their feather fetish? To find out, I deposited a half-dozen each of black and white feathers onto the bare soil in the garden, to see if the swallows had interest. The female flew at least a dozen times over them before she took one. But instead of flying into the nest with it, she took it into the sky, and while circling high, she dropped it. As it slowly drifted down she turned back, caught it, again circled with it, and dropped it again. I then started counting: she continued flying higher and making shorter loops before dropping it, again and again and again. She never missed, but on her tenth drop of the feather, she let it drift on without retrieving it; it descended into the crowns of the trees surrounding the clearing. Feathers were now apparently an object of play, not a necessity.

The pair left and came back together several times during the day. At 5:30 p.m. they perched on their usual spot on the locust; so far there was no sign of incubating. I checked and photographed their more than normally feather-lined nest, which effectively shielded their eggs from view. Although it was still possible for the female to lay a sixth egg the next morning, I doubted it would happen, but I could not wait to find out.

On May 27, a swallow *chilp*ed (a sound distinct from monosyllable chirps) high in the still-dark sky at 3:55 a.m. The sound continued unbroken — on and on, without pause. At 4:58 a.m. a different pitch, a chittering, started, and then there were two swallows. They came lower and met at times to hover briefly together, and then separated. The male later again continued *chilp*ing as before, and then as the pair again met in the air, the male sang his song instead, but briefly. After this the couple landed on their perch on the solar panel, and the female made several passes by the nest-box before entering it at 5:18 a.m., as he resumed his sky dance.

When the sun finally lit up the treetops at 5:25 a.m., the male was still circling and *chilp*ing, and so he continued even at 5:50 a.m., as the female remained in the nest-box, either incubating or perhaps laying a sixth egg. But when she left the nest-box at 11:05 a.m., the nest still had only five eggs.

And after this day, with the last egg laid, the male's predawn sky-dance displays stopped completely. Their synchrony with her egg laying remained for me a mystery.

Tree swallow eggs have a variable hatching time, reportedly between twelve and twenty days from laying, presumably depending on incubation time and temperature. Both would vary, depending on how quickly the female can feed herself each time she leaves the nest. In this nest, five baby swallows hatched on June 10–11. Sixteen days of incubation had passed uneventfully, except for some days of cold and rain. The third swallow had only seldom come near during that period, but the pair at least once each day flew around together, and the male often sang quietly when perched on the panel after the female had entered the box. They often took turns at the nest.

Curiously, when I finally again examined the nest-box on June 17, it contained only four babies, and in the nest there were no eggshells nor an unhatched egg.

By June 27 the four babies were in black stubble, starting to feather out. The weather had turned sunny and warm. The babies were quiet in the nest. None were hungry enough to come up to perch in the nest-box entrance to intercept food; the parents were entering the nest at each feeding. I therefore expected this time to see the whole clutch fledge together because there had been no competition among them to intercept the incoming food at the entrance. Given the reduction in the number of nestlings, I expected the fledging to be a success.

On June 29, the female attacked me when I went near the nest-box. This had not happened throughout the whole nesting period. Now she flew high into the air, then dive-bombed straight down toward my head and face, at an estimated forty miles per hour, missing me by less than ten centimeters. Then she rose into the air to do the same again, and again; fledging of the young was clearly imminent. Whether conscious of it or not, the parents anticipated their young's great vulnerability by becoming defensive.

On the next forenoon the swallows occasionally came to the nest entrance, and sometimes the young cheeped when a parent flew by. I had to be briefly absent later in the afternoon, and when I returned (in a hurry) at nearly 3 p.m. there was silence. There were no young in the nest-box. However, after a few minutes a swallow came and chirped. A reply of cheeping came from a tree where none of the swallows had landed before. It was one of the babies, and it flew out of the tree and joined the adult in flight. The two flew off together in the direction the parents had often taken for their foraging trips. I had missed the fledging by perhaps minutes, but one of the young had lingered behind as the first three left. The parent that had returned to retrieve this baby knew that it was the last of the clutch; no swallow came back to this clearing — not that afternoon, the next day, or most likely the entire summer.

The nest-box, unlike most, retained a dry snug nest. It did not contain a compressed mat of moist guano. The young had been well fed, and none had hogged the nest-box entrance. The parents had been able to enter not only to feed the young but also to remove their waste, which was excreted in easy-to-carry sacs upon the stimulus of the parents' arrival at feeding times. No parasite egg had appeared; only one egg per day had been added, and one nestling reduction had occurred. The fledging of four healthy young is a successful nesting.

Day-old hatchling, unhatched egg, and fecal sac of a full-size baby.

9

SERIAL TENANCY AT THE NEST-BOX — 2016

DURING THE MONTHS AFTER FLEDGING, TREE SWALLOWS assume a new life. They become attracted to one another, first keeping company with dozens of other swallows, then hundreds, and then thousands. As winter arrives, Maine swallows inhabit first the coastal marshes along the northeast Atlantic coast, then move on to those of the southern New England states, and next continue on to the Carolinas, Florida, Louisiana, and points even farther south, in Central America, where they swarm in the millions.

Near the end of winter, the swallows take the several-thousand-mile return trip in stages, but this time at a faster pace. As the throngs of swallows move northward, groups split off to specific home areas. From one or another such groups, birds then depart for their previous nest sites. I followed them, with pleasure, but only in my imagination and through the knowledge I had gleaned from others' writing over the past century.

Anticipation for the swallows' return had been building up throughout snowy, cold March. I was eager to see them again as individuals pursuing their lives in circumstances that might change in a matter of seconds. There was little way of knowing what might happen in the next year, and thus much reason to watch again. I hoped to determine which of the observed patterns of behaviors were typical in terms of an evolved reproductive strategy, or merely incidental.

AT THE END of the first week of April 2016, a vernal pool at the edge of the clearing was still covered by thick ice. The woodcock was back, and on April 9, a flock of seventeen male cowbirds dropped in to feed on corn left out for wild turkeys. The first robin was singing, and a sapsucker was already making a sap lick in a sugar maple. Temperatures dropped from a high of 40 degrees F to below freezing overnight, and it then snowed.

On the 16th, the warmest day so far, the first phoebe arrived and immediately inspected potential nest sites at the log cabin. The ice of the vernal pool had melted, and wood frogs were chorusing. A flock of tree swallows had been spotted down by the lake, and I got up early the next day, anticipating that I might see one or more in the clearing. The brook was rushing loudly from the spring melt, a mourning dove cooed, and the first hermit thrush was singing from the edge of the clearing. By 5:30 a.m. a chickadee chimed in with its *fee-bee*s as the sky lightened to pinks and yellows. The first mourning cloak butterflies were out from hibernation, and I expected swallows. As if on cue, I heard several cheeps, and saw one swallow circle and land on the black locust tree, on the specific branch, and the specific place on the branch, that the pair from the past season had used. It perched there silently for several minutes, then left.

It had started raining on the evening of April 18, and then it poured through the night, turning to snow by morning. I saw no more swallows. My birthday, April 19, was noted with a home-made blueberry pie. The next day temperatures rose to 40 degrees F, warm enough for flies to fly, and at 8:30 a.m., with loud cheeping, three swallows arrived. One immediately landed at the entrance of last year's nest-box, at the edge of the garden. Together with another bird, it entered unhesitatingly, as though the two owned it, and they probably had. One of the pair then flew to another nest-box, which had also been used by swallows in earlier years, before flying to perch on and sing from the solar panel, another favorite perch of the two birds last spring.

The three included a pair, because whenever the third swallow flew anywhere near one of those boxes, the male returned to the nest entrance to perch and block it. Most of the pair's attention was focused on the previously used box, and the male flew to the other box only when the third swallow showed interest in it by coming near. As he had done last year, the male of the pair ignored me when I stood by him at the solar panel, where he sang.

Temperatures were barely above freezing on April 21. I was up, to await the swallows as the horizon was lightening, at 5:15 a.m. The hermit thrush sang, the phoebe called, a robin sang, sapsuckers drummed, but no swallow arrived until an hour later, when the sun had started to peek over the ridge. It was a single bird and it landed on the locust tree, preened, and stayed silent. Ten minutes later, it cheeped, left its perch, circled, and flew out of sight. It returned at 6:45 a.m., was silent again, landed on the same place on the locust, preened, and in two minutes started circling high. Again it returned, at 7:25 a.m., stayed silent, and left after twenty minutes. A minute later three swallows arrived, and one of them swooped down to last year's used nest-box by the garden, then left. The same pattern — only one swallow staying, remaining mostly silent, and then leaving, followed by a pair arriving, with one of them checking the nest-box — continued unchanged over the next four days.

On April 26, the whole eastern sky was red at dawn before it snowed. No swallows came. But the next dawn, at 20 degrees F, one swallow arrived ten minutes after sunrise and continued the come-and-go pattern, as before. But at 7:20 a.m., after this bird had perched quietly, it suddenly started to chitter excitedly, and then it circled swiftly into the clear sky, to join a pair there. The three appeared to play together, sometimes soaring, sometimes flying alongside one another. Gradually they drifted apart, and a minute later the lone swallow returned to its perch on the locust tree, then left at 8:11 a.m.

April 29 dawned cold (20ºF) and clear. No swallows were present at sunrise (5:52 a.m.), but one arrived at 6:50 a.m. It circled briefly and landed on the familiar spot on the locust, and left quickly. It promised to be another routine day, and I went in for breakfast and did not come back out until a half-hour later, when the sun was high. A male swallow was then perched on the solar panel. I walked over to him, and he paid me no attention. He just kept on preening, stretching, and singing. He sang at a very low volume; therefore he was not singing to attract a mate. Knowing he would not, however, be singing idly for himself, I looked at the nest-box and, as expected, I saw a female peeking out; he had been singing to her. On this and the next day the two came and left repeatedly, and a third joined them once as they soared.

Though I did not understand the pair's repeated interactions with the unidentified third swallow, one thing was certain: this pair of swallows had come to an agreement regarding their nest-box. As I observed their protracted interactions, it seemed likely, based on his behavior and appearance, that the male was the one from the previous year. The female was a new mate, one molted into the more mature plumage. She had more of a silver sheen than most, and except for the head, which looked less metallic, she was difficult to distinguish from the male. They were now defending their box, and not just against other swallows. At 8:17 a.m., when both were perched within ten centimeters of each other on the panel, a chickadee entered their nest-box. Both swallows then fluttered in front of the box, and the potential usurper popped back out and left. The pair then convened on the black locust tree. Then the female again fluttered down to the box but did not land, and then flew to the panel, while the male remained on the locust and sang. She then joined him, and he sang softly in his usual long monologue of chirps and cheeps, terminated by a crescendo of throaty gurgling-tinkling, with a louder downward inflection at the end. On and on he sang, as she stayed silent. After one of these episodes of singing, he suddenly made excited *chee chee chee* calls and chased after a third swallow, while the female remained on the box. The third swallow appeared twice more. The male of the pair again made *chee* calls as the third swallow swooped over the nest-box, but he did not attack it. Instead he escorted it off, then returned to the solar panel.

He then acted as if he were trying to induce his female to enter the box — flying to the entrance, clinging to it, and singing there whenever she came near. When she didn't enter, he then went in himself and sang from the inside. When she finally did slip in, he immediately exited and sang from the nearby solar panel, as she peeked out from the entrance.

THESE SWALLOWS WERE fortunate to get here in time to choose a nest-box, and to find a mate, or be reunited with a mate, to potentially defend and use this shelter. Nesting cavities are usually limited in quantity, and numerous other bird species use them as well. They include bluebirds, house wrens, European starlings, English sparrows, great crested flycatchers, tufted titmice, and chickadees. In Maine the native black-capped chickadees could have gotten a head start on these swallows in the competition for such a nest site. The chickadees are year-round residents and can forage in the forest all winter long, ready to start nesting before any insects are present for the swallows to feed on.

Black-capped chickadees had in prior years used this nest-box, and one year their nest had contained both swallow and chickadee eggs at the same time; only the swallows' young had fledged. This year as well, chickadees got to this box first and had started building a nest. But that did not deter the swallows from claiming it when they finally arrived. One or the other of the chickadees now kept revisiting this box, and the female swallow would attack it in the air. In the tussle the entangled birds tumbled, fluttering, to the ground; these furious bouts were like those between swallows that I've previously described. After one such encounter, this female swallow chased the chickadee into the bushes, a habitat that swallows usually do not enter. The male swallow, meanwhile, merely fluffed out his feathers, a behavior unlike any I had seen before under similar circumstances. The female swallow continued to defend the nest when the chickadee came back, and this time she chased it far into the dense woods. I didn't see her come back out, but when I went to check, I found her perching silently three meters above the ground, among thick branches in the bushes, her head rapidly swiveling all around. She stayed in there for ten minutes.

Over the next five days (May 1–5) the weather shifted to rain and then, atypically for this late in the season, temperatures dropped and it snowed. Such weather favors chickadees because they are tree gleaners. They don't depend on flying insects for their food. The snow had chased the swallows away, so the chickadees were able to resume work on their half-finished nest in their box — in fact they almost finished it.

On May 6, the temperature rose to 45 degrees F by 7 a.m., under a clearing sky. The pair of swallows had returned by 7:10 a.m. and remained most of the day. Four times, a third swallow appeared, and the three circled high and flew peacefully together. Later, the temperature climbed to a balmy 68 degrees F, triggering a large hatch of black flies: Maine's unofficial first day of spring had arrived. The day was also a marathon watch for me. From dawn until 4:20 p.m., I jotted down approximately 410 behavioral acts documenting the renewed battle between the swallows and the chickadees. The swallows won (again).

At 3:33 p.m. on May 7, as both swallows were positioned directly in front of me, I pulled a feather from my pocket and tossed it to them. What happened next threw me: the male did not fly to me to get it; instead, he flew to the female, fluttering at her and then landing beside her. As I expected, at this time before nesting had begun, both ignored the feather. However, *she* then flew at *him*, and they both left. It looked as though he had tried to induce her to get it, and she had said, "No, it's yours, you get the feather." But neither was yet motivated to get the feather for itself. They were perhaps not yet convinced they would have a chance to nest here or were not in a playful mood.

Not much happened over the next three days, except for the flyby of a merlin. The pair of swallows, perched in silence on their locust tree in the fog of a rainy, cool morning, made loud *chee chee chee* alarm calls and launched to chase this small falcon, which is about the size of a blue jay or a mourning dove, birds that the swallows ignore totally. But then a chickadee flitted into the clearing, and the female swallow instantly attacked it.

Chickadees never attack swallows. They don't have to. They normally nest deep in the woods, at almost any elevation, from close to the ground to ten meters up. And if they don't find a suitable premade cavity, they hammer one out on their own in a rotting tree trunk. This pair of chickadees would have saved themselves a lot of work if they had preempted the swallows in securing the nest-box in the clearing. This year, though, neither the chickadees nor the swallows got a real head start.

Black-capped chickadee eggs in a tree swallow clutch. (June 6, 2007)

Swallow swooping to take a white feather from my hand. (June 3, 2014)

Babies hatching. (June 12, 2014)

The young seem ready to fledge. (July 1, 2014)

The same young, dead the next morning. The beetle was with them. (July 2, 2014)

Male swallow scanning the sky from the nest-box entrance. (June 7, 2016)

Female swallow, with a fresh spot of paint, in the nest. (June 11, 2016)

Male at the nest-box entrance. (April 28, 2017)

Four fresh eggs and
feathers in a nest.
(June 6, 2017)

The same nest with a full clutch.
(June 15, 2017)

Four hatchlings and one egg.
(June 15, 2017)

The male
on the nest.
(June 21,
2017)

The female on the same nest. (June 22, 2017)

Nest with four half-grown babies (only two visible) and two eggs. (July 2, 2017)

Four nearly grown nestlings. (July 13, 2017)

Two remaining babies.
(July 15, 2017)

The last one out, still too weak
to fly. (July 17, 2017)

A barn swallow nest with five young. All are nearly the same size.
There are few feathers in the nest. (June 20, 2016)

The same nest, with a parent feeding a chick. Note that all have access to the parent bringing food. (June 20, 2016)

Same nest, with all the young nearly equally satiated. (June 20, 2016)

The apple-tree male on his favorite perch. (May 19, 2018)

The apple-tree male in the nest-box, with half-grown young. (June 30, 2018)

His mate at their nest-box at the edge of the garden. (May 19, 2018)

The beaver bog within two kilometers of the clearing. In these trees, killed by the water, woodpeckers left nest holes that swallows can use. (January 28, 2018)

The clearing with locations of nest-boxes indicated by X's, with numbers referring to years used as nesting sites, and the three trees used as perches (BL=black locust, A=apple, M=maple).

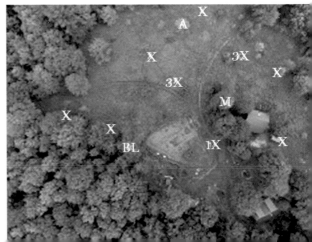

10

MATING TIME — 2016

ACADEMIC RESEARCH CONCERNING TREE SWALLOWS IS dominated by a single topic: extra-pair copulation. It has been the focus of more than sixty scientific publications. I am fairly sure, though, that the mating I saw, and I saw a lot of it, involved just the pair of swallows at the nest site and no others. I was keeping an eye on both the male and the female, and I saw them chase intruders away. This did not, however, preclude extra-pair copulation offsite. If this did occur, it likely involved a cost to the perpetrators because vigilance in guarding territory and the nest itself would be compromised. The risk of another pair's egg being cuckolded into the nest-box would increase. And the cost of a cuckolding could be huge. I therefore suspect that for a pair of swallows in a natural clearing, the male and the female would both tend to stay to defend their nest instead of leaving to seek extra-pair mating. It may be a different story for the many birds kept at study sites, in grid boxes. The swallows I observed were more isolated.

Like all swallows, tree swallows share the adaptation of gleaning insects as they fly swiftly over open areas; they cannot forage in the forest. For millions of years, beavers have made openings in the forest, providing ample airspace. Beavers flood these forest areas, creating habitat for aquatic insects — abundant food for swallows — and leaving dead trees, with old nest holes made by woodpeckers, usable as nest sites. Key to understanding tree swallows is, I believe, their nesting in tree cavities in food-rich areas near or over water, where the larvae of many flying insects overwinter, to emerge in spring and summer. The swallows' adaptation to woodland, to wet insect-rich clearings, and to used woodpecker holes meant that nest holes became a limiting factor in their reproduction.

Nesting in tree holes would have evolved highly aggressive behavior toward other birds, especially other tree swallows, and the premium real estate went to the swallows that returned earliest from migration. Woodpecker holes persist for many years, and knowledge of a great nest site would promote nest-site fidelity and the reunification of mates, who would learn how to cooperate as parents and as defenders of a nest site. As predicted by this scenario, the tree swallows in my clearing were intensely aggressive in protecting their turf. There are, however, consequences associated with such nest-site fidelity and territorial defense.

It is not a sure thing that a swallow's mate will show up again the next spring. There is high annual mortality among tree swallows. A "backup mate" for nest guarding could be important. But it's possible that, later on, the missing mate may show up after all. The new mate has invested time and energy in the nest site and would gain little or no advantage from having the previous mate nearby. Yet the new mate may not be motivated to be aggressive toward the former one. If the latecomer is the male's previous mate, might he tolerate her? Or might she cuckold the nest with offspring difficult to distinguish because it shares genetic markers with the other young? (If the latecomer is a male, the female might tolerate him at her nest, but she would likely already be mated. Generally, in any case, males are usually the first, not the last, to arrive.)

In addition to these complexities, what are the options for young birds that fledged during the previous season if all the real estate is claimed before they arrive? They have no "squatters' rights" over a nest from a previous year. For a young male without a mate, one option is extra-pair copulation. This has little cost for a mated female at a nest — she is not giving up anything. But for her mate, the male of an established pair, the consequences could be serious — he may lose a little or a lot of his genetic contribution to the chicks in his nest. He must be aggressive toward the first-year male. Wasting his effort on rearing another male's offspring is an evolutionary dead end for him.

If the "third party" is a late-arriving female, she could, as we've seen, cuckold the other female by dumping her own eggs into the nest. The mated female might tolerate the male's extra-pair copulation if her own egg contribution to the nest is secure. But she cannot risk having another female dump an egg into her nest. The reason for this is clear: the cost of having too many young is high. An extra egg that becomes an extra nestling could endanger the entire clutch. Starvation of the young is the penalty. And since a female has only a fifty-fifty chance of surviving to the next nesting season, this one clutch potentially represents her lifetime reproductive effort.

The swallows that in the spring failed to get a nest site would have an interest in others' successful nests, and the owners of these nests might have reason to welcome them. These non-nesting swallows may be scoping out potential nesting options for the future and also a potential mate ready to capitalize on an opportunity to claim and hold a nest then. As for the resident pair, a possible future benefit might make newcomers tolerable. One of the pair might fail to return the following spring; in that case, a swallow already familiar with the location could become a new mate. My observations were not fine enough to determine which of these possibilities informed particular behaviors at the nest-box in the clearing. But the distinct differences in the resident pair's tolerance levels for different birds suggest that these possibilities do exist.

ON MAY 11, the barred owl's song, *who cooks for you,* repeated again and again, a turkey tom was gobbling nearby in the still-dark woods, and ovenbirds were singing in the maple groves long before the first swallows appeared. Three were finally circling at 5:35 a.m. as the sun was starting to come up over the hills, and after ten minutes the pair landed and the other swallow left. My note taking took on a furious pace as I tried to keep up with events on this day when the birds' interest in nest building, and feathers, was clear. It started as the male flew to the nest-box, perched on it, and sang. The female came down to it also, ducked her head in several times, entered, came back out, and flew close over the recently tilled garden, for the first time showing interest in nesting material. After that thus far unusual flight, she perched on the solar panel, where he was singing, and he hovered over her briefly, in what looked like copulation, but with no contact. She then flew into the garden, to land on the bare soil. Three times she did this, but on the fourth landing she finally picked up a stray stem of grass. Her first! She flew up with it, but not to the nest-box. Instead, she returned to the male far up in the locust tree, where he was singing. And there she dropped it but afterward flew repeatedly to the same open ground in the garden. Sometimes she merely flew up again, and sometimes she picked something up and then dropped it before coming back to the male swallow in the locust tree. However, by 9 a.m. she was regularly bringing grass into the box and quickly reemerging to get another blade. He stayed perched and sang the whole time.

Knowing from previous years that the male was the main provider of feathers, it seemed odd that he hadn't budged yet to find any. Maybe she would not yet accept them? To find out, I dropped a long white duck feather onto the ground where she had been gathering grass. She instantly swooped over it but did not pick it up. I then tossed it into the air. This time she picked it up off the ground after it landed, flew several times around the clearing while holding it in her bill, and then tried entering the nest-box with it; but it was much longer than the entrance hole's width, and she was holding the feather cross-wise in her bill. She tried again and again to enter but eventually dropped the feather. Several minutes later she retrieved it and flew around with it, only to drop it three times. But she caught it only twice, so on the third drop it drifted into the weeds. A half-hour later she picked up off the ground an even longer but more flexible feather, circled high over the clearing with it, dropped it, and made five drops and catches in a row. This all took place in front of the male, who remained perched on the locust tree.

There were still only several grass stems in the developing nest (on top of the now long-abandoned chickadee nest). But this female had shown, as before, that female swallows are not only willing to pick up feathers but also capable of doing so. Why had the male of the pair again ignored the feather, when, as his previous behavior suggested, he was the main feather provider? It was a critical time to observe details.

The next day (May 12) started with a blazing orange sunrise, as the pair of swallows arrived at 5:26 a.m. and perched side by side, as usual, on their locust tree. Suddenly, after nine minutes of peace and quiet, one swallow was pursuing another in a wild chase, but not in the sky, as was typical. Instead, the two wove an erratic course through the trees, banking and weaving. The chase was much too fast for me to assess who was chasing whom, and I had no clue what this behavior might mean. But within less than a minute the female was again perched on her spot in the locust tree, the male was fluttering over her back, and they mated. It was the first consummated mating I had seen this spring. The intensity of the chase immediately previous to it seemed odd, as did the vocal silence during the flight. There had been no alarm calls, and I had not seen a third swallow come near.

At 6:30 a.m. the female picked up the first grass stem of the day, circling with it three times before flying to her mate and perching beside him, with the stem in her bill. Two minutes passed before she finally slipped down into the nest-box, still carrying the grass, to perch in the entrance and peek out. Five minutes later she picked up her second piece of grass, flew into the nest, and then again peeked out. The male, meanwhile, re-

mained on his perch and sang. By 7:05 a.m. she had made six
grass-collecting trips. On one she dropped her straw on the
way but entered the box empty-billed anyway, and that time
she stayed inside for thirteen minutes before leaving with her
mate for a two-minute flight. After that the two landed on the
locust tree and mated again, and then she provisioned the nest
with one more piece of grass. Again she returned to him, and
they mated eight more times before she returned to the box and
briefly stayed inside.

The start of mating seemed an appropriate time to check for
any behavioral link between mating and the previously observed
feather fetish of the male. He soon came close to me, perching on and singing from the panel, and I tossed two large white
feathers onto the ground. Both birds swooped down. The female
took one and carried it into the box. The male took the other but
dropped it in flight, and it fell into the weeds in the field. When
I repeated the trial with two new feathers, each bird again took
one. She took hers into the nest-box, and he again dropped his.

The swallows left the clearing at 9:45 a.m. and, shortly after their return twenty minutes later, another violent and silent
chase, like the previous one, took place. It again happened within
the space of a moment: all of a sudden two swallows zoomed by,
so close to me that I heard the zipping sound of rapid wing-beats
as they swerved, evading each other. I was again unable to identify the participants, but afterward the male flew up to the perch
on the black locust, where the female joined him, and they mated
twelve times; then, within five minutes she made three trips to
the garden to gather grass for the nest. Since again I saw no signs
of a visitor, and since swallow mating appears to involve a del-
icate balancing act wherein the male appears to stay in flight,

hovering up and down over the female and touching down only for the precise but momentary cloacal kiss, my best guess is that the chase was a form of play that discharged tensions.

IT WAS MAY 13. A swallow was cheeping, flying high above the clearing at dawn. It was alone and continued to circle, but at 5:40 a.m. a second swallow joined it. In two minutes both dived down, in silence, to the usual perch on the locust tree, and there they started shaking and preening themselves. After twelve minutes they both returned to the sky, soaring in short intervals; this was unlike their earlier flight, which was continuous and fluttering. They left, came back, left again, returned, then left for nearly two hours. The black flies were killing me. It was an agony to try to keep track of the swallows' behavior. But I had to, feeling that something unexpected had to happen soon, as it usually did.

At 9:05 a.m. the male returned to his perch on the locust tree, and the female went into the nest-box and peeked out. "Another good moment for a feather," I thought, and waved a white one in the air. She flew toward me as I tossed it, and it landed on the ground. She picked it up, flew two circles around the clearing with it, and took it into the box, and then again peeked out. The male did not budge from his perch. Both then left together, returned at 9:45 a.m., and mated five times. Then, pronto, they left and did not return until noon.

Rain poured down that night, but in the morning of May 14 the sky cleared, and at 5:05 a.m. there came from on high a nonstop chirping. It continued until 5:30 a.m., when I first saw the swallow. For some time it remained at a height of about 120 meters, too high for me to follow it by eye. Ten minutes later I saw two swallows, and they were lower. Then, at 6 a.m. a third joined the pair, and I heard excited calls of *chee chee chee* instead of the usual chirping. The three circled together, in perhaps token chases, but they showed no obvious signs of antagonism. They at times flew together until 6:02 a.m., when they separated but circled on and on in silence, in a sky of scattered wispy clouds. The pair descended to their perch at 6:12 a.m. and mated nine times in succession, then perched side by side. Again they mated, three times, and after a short flight they came back again for five more matings. The third bird returned, and the three again circled together, but this time only for a minute. After this the male sang, flew to the nest-box, and peeked in, and the pair circled together. They landed at the same spot and mated four more times.

There was still no egg in the nest-box, but the time for egg laying was at hand. If a third swallow wanted to sneak its own egg into the nest, now would be just about the right time — but not quite yet. This maneuver would likely be unsuccessful at this precise moment because the resident female, not yet having an egg of her own, would likely recognize a parasite egg as a stranger's egg, even if that egg resembled a swallow's egg. To test this theory, I tried an experiment. The phoebes were now laying their eggs, which are also white, and I used one of theirs to fake a swallow-nest cuckolding, placing the phoebe's egg in the still-empty nest of the two swallows.

Twenty-three minutes later, when the pair returned to the solar panel, the male flew into the nest-box and then back to the panel, where he sang. The female then went into the box, but quickly popped back up to the entrance, hung there, and repeatedly kept looking in. Five more times she repeated this behavior, going partway into the box and coming back out, until she finally ventured all the way in and stayed down. She remained silent the whole time. I had not seen such behavior before and suspected she would very soon lay her first egg. She stayed on the nest for fourteen minutes, longer than ever before. But when I checked, no egg had been laid, nor had the swallow removed the phoebe's egg. So I removed it myself, in case it might discourage further nesting activity.

THE TREES LEAFED out over the next three days (May 15–17), rain poured, and a strong northwest wind under a dark sky dropped the temperature to near freezing. The phoebe and the chickadees still sang, but no swallows came until the 17th — when at 6:30 a.m. it warmed to 38 degrees F, and I saw two swallows fluttering in fast flight, high in the sky. They could not have found insects here over the past two days. But they could have flown 150 miles, in five hours, to find food. Their world is larger than a phoebe's, a wren's, and many of ours.

By noon it had warmed and the black flies were out in bloodthirsty hordes. Swallow watching became torture. There were still no new feathers (I had removed the two I had offered earlier), no further signs of more nest building, and no egg.

On May 18, the faint *chilp, chilp, chilp* of a swallow sounded from the puffy clouds high above, forty minutes before the sun came up over the horizon. Between the clouds and against the sky, the bird appeared as a mere dot, and I had to strain my eyes to see it as it disappeared from view, then reappeared. At times I could follow it only by sound. As the sun started to crest the horizon at 5:28 a.m., this single swallow continued circling and calling, without a break. Then at 5:34 a.m. there were suddenly three swallows flying together, and six minutes later the resident pair landed on the usual perch on the black locust and immediately mated. They perched within five centimeters of each other, and the male sang; then by 7 a.m., when the clouds had cleared and the sun shone brightly, the male landed on the panel and later slipped into the nest-box, where he sang. The female immediately flew down to perch on the post by the box, and he returned to the panel and sang there.

The nest was now finished. The female herself had gathered all the grass, which had all come from the tilled garden plot. But her next moves seemed puzzling: she again flew to the ground, picked up a piece of grass, and flew to the nest entrance, but this time, rather than take it in, she brought it to her mate, dropped it, and flew back and into the box. He sang, and she immediately popped back up to the entrance. She again flew out to pick up a piece of grass, which she again brought to him and then again dropped before returning to the nest. And then she repeated the sequence a third time. Was she careless? Or could she be giving him some signal, perhaps trying to nudge him to do his part — perhaps urging him to bring feathers?

At 7:45 a.m. she once again fetched grass and brought it to the male as he perched on the panel, and then she dropped it. He sang. She flew back to him and perched beside him, he fluttered above her, and they mated twice. Then she finally brought a small piece of grass to the nest-box but hesitated with it and sang once as she hung at the entrance before entering.

By 8 a.m. the two had mated six more times. A third swallow had come, and the pair flew off with it without any apparent fuss. Upon their return, there was only one mating and two apparently halfhearted feints. With mating apparently winding down, egg laying was sure to come soon. But by that evening, the nest was still empty, with no egg or feathers.

The next morning (May 19) I jumped out of bed and was out the door into the dark at 4:20 a.m. because through the open window, in my half-sleep, I'd heard barely audible calls of *chilk, chilk, chilk*. It was far too dark for me to see, and the swallow far too high anyway, but I needed to follow it, and I could, because it called continuously and could be tracked by sound as it coursed the sky above the clearing. And there was only one. One hour later, at 5:20 a.m., the swallow was still circling but had become visible, alternately fluttering and soaring, with periods of silence.

The female arrived at 5:45 a.m. and landed on the locust. The male then came down from on high to land beside her, and they both took off in what looked like a playful chase. Next they perched side by side and mated six times, and then eleven more times (in four episodes). Finally, at 10:05 a.m. she flew to the nest-box and perched in the entrance as he sang, and she entered and peeked out. Then they left.

The by now familiar call pattern of *chilp, chilp, chilp* resumed on the next day (May 20) in the dark hours before dawn. The calls were rapid and rhythmic, fifteen of them every ten seconds, and thus they continued nearly nonstop until 5:20 a.m., when there was a moment of silence. Maybe the swallow was out of range. But at 5:55 a.m. it became visible in the sky, and in five minutes a second joined it. And then a third showed up. Again the resident pair landed on their perch on the locust tree, and then immediately mated, in bouts of two, four, and then six times over the next seven minutes. Then they both flew to the nest-box, where the male perched at the entrance and the female on the post next to it before he then flew to the panel and sang. Finally, both entered the box but then returned to their tree perch, to mate ten more times, in two sessions. Again, it seemed that mating was winding down and egg laying about to commence.

A third swallow flew by at 6:37 a.m., and the pair cheeped in alarm, then mated three times after it had left. The male then sang from the panel and flew back to their tree, where they mated three more times. After that, the female flew down to the post by the box, where the pair mated four times before returning to the panel. She now picked up a piece of grass, he sang, and she then carried the grass into the box. He entered it as well, and quick-paced whispering-singing came from inside before he flew out and sang from the panel. She stayed silent in the box, then peeked out for four minutes before flying out to join him, and he immediately stopped his singing. Both left briefly (for six minutes) but returned to the tree, where they mated eight times. No other swallow was near. Three times she picked up, and finally carried, a piece of grass into the box. The nest now had a round cuplike depression, and except for the fact that there were no feathers, it looked finished. The female stayed inside for two minutes, then left to land directly in front of me — an unprecedented gesture.

Why should she land by me? Was it a random act? Maybe, but then I had a random thought — might she want a feather? Frustrated that her mate had brought none, was she now soliciting one from me? It seemed unlikely, and I would never know. But I had indeed provided feathers to her in the past, and perhaps she associated me with them, as she might associate them with her mate also. I ran into the cabin to find my feather cache, grabbed a white feather, came back out, and tossed it into the air — she instantly swooped over to me but did not take the feather as it landed on the ground. She swooped over it five more times while the male swallow perched nearby, in plain view, on the panel; he sang. Finally, I dropped the feather in front of *him*, though he must have seen it already, along with his mate's behavior. He ignored the feather, but she then flew over to it, picked it up, and carried it into the box. Was I seeing correctly? This seemed incredible. I repeated the sequence — and got the same result. And again she perched near me. Was she training me now, because she had failed to influence him? The idea seemed preposterous. So I did it again; I tossed her a feather, then another — and each time after she carried the feather into the box, then perched on the garden post next to me. The behavior was starting to look like a pattern, and I tried still another white plumose feather — same result. I tried twice with black feathers. She ignored them. I tried a small, curled, fluffy white feather; she ignored it too. The male still hadn't budged from the panel. After this I tried once more with a large, flat white feather. This time, after making three passes over it, the female picked it up, but instead of taking it into the box, she flew high into the sky with it and dropped it four times, retrieving it each time, only to drop it again.

A third swallow arrived at 9:20 a.m. The pair flew up to meet it, and they circled side by side, before the pair returned and perched on the panel. The female then returned to the nest-box. When she peeked out, I dropped two white feathers in front of her (and within four meters of the male). She took both feathers into the nest, after which he flew to the nest entrance, then back to the panel, to sing. It looked as though he had now at least acknowledged her efforts or was acting to encourage her to do more.

She now perched in the nest-box entrance, then flew to the pole. He joined her, and they mated six times, after which they left together. Was mating a distraction to him now, one that kept him from hunting for feathers?

When the two returned an hour later, she again perched near me on the garden fence post. The pair made several flights together over the next hour, and they seemed at ease, until 10:48 a.m. Then he fluttered twice over her back, in what looked like an attempt at mating. But this time there was no contact; she didn't scooch down into the proper posture. A frenzied silent chase ensued through the trees. It lasted less than thirty seconds. As was the case earlier, when I had observed similar frenzied low-altitude chases, I had not seen a third swallow; but, as before, I had not seen the first seconds of the flight and therefore could not know if it had involved only the pair or also a secretive newcomer. Immediately after the chase, the pair resumed their perch on the locust tree and mated there several times. They then perched side by side.

On May 21, mating continued, and the female brought a white feather into the nest. It was the first one that I had not provided. A third swallow appeared three times, but there were no chases.

It was seemingly night when I woke on May 22 at 3 a.m., hearing calls of *chilp, chilp, chilp, cheep, chilp, cheep* through the open windows. It was either the earliest arrival of a swallow so far or my earliest wake-up to notice, and so I leapt up and ran out. No doubt about it: this was real, as I had observed routinely. I counted the metronomic sounds at eighteen per ten seconds, and stayed put and took notes for the next five hours without a break, forcing myself to hang on to the end so I could see how long this aerial display could last. At 5 a.m. the swallows were silent for a while and were not flying at all-out speed as before, but they were still soaring. But finally at 5:40 a.m. the pair landed on their perch on the locust, and they mated seven times. Five minutes later they mated eight times more, and she then flew down and entered the box, while he remained perched in the tree and sang until she had entered. Then, after ten minutes, he left, and she remained alone inside the box for fifty-five minutes. When she finally departed, I thought she might now have laid an egg at last, but when I checked the box a half-hour later, when she was again in the sky fluttering, silent, and soaring, the nest was still empty.

There was still no egg by the end of the day. What were they waiting for? Why all the daily chirping and flying in the dark? My patience was tested by the now constant assault of black flies, which started soon after sunrise. But I could not stop. Many dots are needed before you can connect them into a picture of what it is like to be a tree swallow. Through these birds I had a view into a world shaped by the inexorable forces of evolution, for producing whatever enhances reproduction, no matter how cruel it may appear. We are not uniquely different, set apart from all the millions of other species on this planet, except for this one characteristic: we are capable of escaping evolution's invisible grip because we are beginning to understand how it works, and we can block it. And here on view, in all its details in the wild, was a demonstration of how evolution does its work in a creature that is quite similar to us, compared with all the others. The swallows suggest paths our own behaviors might take when unconscious emotions control us.

And so, on this day, with no egg yet in the nest, I continued collecting more data. The female swallow spent the rest of the day in the nest-box or in its entrance. The male came and left, and they mated at least thirteen times. No nest material was added. Surely the first egg would be laid tomorrow, I thought. Perhaps the vocal sky alarms I had observed were meant to foil any attempt to insert a cuckold egg into the nest. But the swallows had no way to prevent me from dumping in a fake parasite egg. The phoebes had not yet hatched in the nest under the cabin roof, and I once more borrowed one of their eggs and placed it in the swallows' nest. (I was a good neighbor — after my first such experiment I had returned the egg to the phoebes' nest.)

The swallows returned at 7:20 a.m., and the male chittered and hovered over the female as though mating, but these were now mere feints — there was no contact, no touchdown. Immediately after, he flew off, and she followed him off the perch and returned to the nest-box. He then went to the tree and sang. A similar pattern — her flight to the box and his to the locust — continued until the end of my watch, at 4 p.m. They made no noticeable response to the new egg, and no additional eggs had been inserted.

On May 23 I awoke to a full moon and a stunning dawn, with no wind and a clear sky. Again I heard the *chilk, chilk, chilk* of the swallow's sky dance before daylight. Except for the phoebe egg, the nest was still empty, and there was still only one bird flying over the clearing, but its call had changed. It was now voicing *cheat, cheat, cheat*s along with *chilk*s. I took a short breakfast break, and when I came back out at 7 a.m. no swallow was flying. I checked the box and this time found the female inside, on the nest.

She did not budge when I removed the front panel of the box to look in, and she later peeked out, at 7:20 a.m., and stayed in the box entrance for twelve minutes. When she left, she joined a second swallow flying silently, high in the sky, in the distance. I wondered if she had been waiting for the male to come, and then left as soon as she saw him from the nest-box entrance. With both gone, I now checked the box: the first swallow egg had just been laid!

I returned to the cabin, impossibly relieved. The swallows' attempt at nesting had seemed endlessly protracted, due at least in part to the changeable weather. I could finally link the recent strange behaviors — the frenzied chases, the sky-high vocalizing in the dark, the feather play and the feather gathering — to the swallows' cycle of nest building, mating, and egg laying. I collapsed onto the couch, physically drained but filled with satisfaction. A major milestone in the swallows' annual ritual had been passed. It had involved mysterious behaviors that I did not understand, which made them more exciting. I felt relief because I no longer needed to watch continuously, recording events moment by moment as I tried to make sense of a sequence of behaviors or determine who was mating with whom. Now there would be weeks of presumably routine activity — egg laying, incubation, and feeding the young.

11

EGGS AND MORE
FEATHERS — 2016

THE SWALLOWS' SECOND EGG WAS IN THE NEST BY 7:15 A.M.
the next day (May 24). The pair of chickadees, not hampered by
cool weather, had in the meantime built another nest in a neighboring nest-box and had caught up with the swallows: the chickadees had just laid their eggs. Perfect timing: I could use one of
theirs to parasitize the swallows' nest, where it would join the
phoebe egg, which had been accepted and was still present. With
its brown spots and smaller size, the chickadee egg would visually contrast with the swallows' own egg, and it might be noticed
and rejected more easily than the white color-matched phoebe
egg. (If the egg was accepted and hatched, my plan was to return the baby chickadee to the chickadee nest — and ditto for
the phoebe.) For now, the swallows made no commotion; they
showed no obvious sign of noticing the strange egg.

On May 25 I awoke at 3:30 a.m. and all was still. But by
4:30 a.m. I heard the songs of robins, a yellow-throated warbler,

a phoebe, and ovenbirds. A sapsucker drummed, and a coyote howled. But it was not until the gray dawn, at 5:26 a.m., that high in the sky a swallow's calls of *chee, chee, chee* could be heard. They contrasted with the male's previous morning calls of *chilk, chilk*, which had occurred much earlier. The female was in the nest-box. The male was soon silent but continued soaring until 6:15 a.m., when he made a couple of *chilk* calls, which she immediately answered from within the box. He then descended and landed on his perch on the locust; he stayed silent for three minutes before flying into the box. A continuous chittering then ensued within it. The sounds were soft, and I sidled close, pressing my ear against the box to hear better, to listen in on their private conversation. After a while it stopped, and the male peeked out. It was now 6:50 a.m., and, I believed, the third egg had been laid. When both left their nest to perch on the panel, I checked. Indeed, there it was. The birds remained on the panel, where they then mated six times.

The male would not have entered the box to help in the delivery of the egg. He could do nothing about that. But he may have been getting an update on the contents of the nest, allowing him to time his deliveries of feathers. Rather than wait for these deliveries to begin spontaneously, I could provide the male with feathers, to test his response. It could happen now. And it did — even more dramatically than I had thought possible.

The male swallow was peeking out of the nest-box while I was walking, feather in hand, to the chair from which I intended to test him. But he launched and was flying toward me even before I got there, and so I held up the feather, and he snatched it from my hand and flew straight back into the nest-box with it, while the female was perched on it. He then stayed inside for about a half-minute. I repeated this experiment three times, and each time it was *he* who fetched the feather. It was he who took it into the nest-box. It was not a nuptial gift.

Both swallows continued to come and go for the rest of the day, perching on the panel, the nest-box pole, or the box itself, or hanging at the nest-box entrance. If a parasite egg was to be effectively placed in the nest, it had to be now. An egg inserted after incubation was at a disadvantage. Because of its slow start in growth relative to the others in the clutch, its hatchling would become the runt of the litter. But an egg inserted as much as a day before could even the odds. The hatched chick would have an equal chance of fledging, if the young in the nest in fact fledged. Nonetheless, this additional baby could compromise the entire clutch. Given that potential cost, it is expected that the nest own-ers would have evolved behaviors to guard against such an egg insertion. The swallows' almost constant attendance near their nest is strong evidence that egg dumping does occur, but I had no proof that the feathers had anything to do with reducing its likelihood or entirely preventing it.

The pair now seemed wedded to the nest, even though the start of incubation was still days ahead. Even with an incom-plete clutch of only three eggs, the female kept on staying within the nest-box. Such vigilance during this part of the nesting proc-ess is unprecedented relative to other birds. Because the female didn't budge, at times I literally had to shove her aside with my fingers to see if another egg had appeared. In contrast, phoebes, and almost all other wild birds, fly off the nest if anyone merely approaches, much less stops to look at them. Flying away does, however, come with risks. Any temporary absence of the nest owner can become an egg-dumping opportunity for another bird. But though the swallows' attentiveness to the nest seemed ceaseless, they tolerated visits from other individual swallows. Against whom were they protecting the nest?

Up close, I could distinguish the male from the female, but not individuals of the same sex. This was frustrating because I had to rely on possible behavioral differences to recognize individuals. It would be far better to mark the birds in some way, and that meant getting close to them. I could now touch the pair in the nest without flushing them out, but I could not risk capturing them and holding them in my hand. I decided on a different method. I stood in front of the box, opened it, reached in, and gently dabbed a spot of paint on the incubating female, hoping to mark her head, wing, or bill. It worked. I had placed a small red dot on her bill and also a smudge of red on her chest. A minute later, she flew out and joined her mate on the panel. She preened the paint spot on her chest, and he preened himself also. They then stretched, he left, and she followed. All seemed normal. I was relieved. This gamble had paid off. I could resume my observations.

On May 26 the fourth egg was laid by 8 a.m., and three new feathers (two wood duck and one grouse) had been added. The female was still spending much time in the box or perched in the entrance and peeking out, and I now routinely checked the nest — it was always the same female, the one with the red dot on her bill and chest. Both members of the pair at times left for minutes at a time, and on one of their returns to their perch on the locust tree, they mated eight times.

The last (fifth) egg was laid on the morning of the 27th. The foreign eggs had not been rejected; the largest (the phoebes') remained in the center of the swallows' bunch of five, while the smallest (the chickadees') lay at the periphery. A clutch of seven eggs is rare, but not abnormal, for tree swallows. It was time for incubation to begin. But the female was still repeatedly perching in the nest-box entrance. The male was now usually not nearby, and his visits had become infrequent. Was he out searching for feathers, or for extra-pair copulations? In the evening, the female was again peeking out of the box, and the male was nowhere in sight. She held her head very still for nine minutes before slowly settling back down into the nest. The sun was setting, and she spent the night in the nest.

There was no way of knowing where the male was spending the nights, but he came back in the dark at 3:50 a.m. on May 28 — or at least that was when I heard his *chilp*s in the sky. By daylight, at 5:15 a.m., he was still *chilp*ing steadily, fifteen times every ten seconds. At 5:20 a.m., when the female finally peeked out from the entrance, she turned her head to look up and two minutes later flew into the sky. His calls stopped. She zipped around in fast flight; in contrast, his soaring was leisurely. Both then circled lower, and she dived at full speed at the box, swerved within a meter of it, made a steep turn, and circled the entire clearing once more, and then three more times. On her fifth dive at the box, she hit the bull's-eye — the box entrance — and was in. The male remained silent and continued soaring.

Throughout May 29–June 1 the male arrived, as usual, before daylight. But the days were uneventful. The female incubated but spent up to ten minutes at a time at the nest entrance and up to twenty minutes at a time out of the nest. The pair occasionally preened side by side on the panel, and he stayed at the nest-box, either within or at the entrance, until she came back. The female popped out of the nest when the male returned, and vice versa.

He stayed in the nest for as long as twenty-two minutes — that was the longest she stayed away. Only when she came back did he fly into the distance and procure feathers. Seven new ones appeared in the nest. The third swallow no longer visited. It apparently had vanished.

The red paint smudged on the female's breast was fading, and the red dot on her bill had become less visible. I needed to touch up these marks. But this time, when I reached into the nest-box, she evaded my brush, and I had to grasp her, apply the paint dots while holding her, and put her back in. When she next came outside, she dive-bombed me once and then hesitated at the nest entrance, peeking in numerous times before reentering. She had identified me, but apparently perceived that the threat was primarily inside the box. Nevertheless, when I offered her a white feather, she picked it up and brought it into the box. The male remained calm and allowed me to get close.

I removed all the feathers from the nest on June 1, to find out if the birds would be willing and able to replace them on their own. They found many without any help from me, and most of the feathers were from waterfowl. By June 5 the nest contained so many feathers that the bird sitting on the eggs would be hidden from anyone looking in from above. The chickadees' egg still remained at the edge of the egg cluster, and the phoebes' egg in the center of it.

Because the female tree swallow does all the incubation, one might suppose that the male would support her by feeding her; this is the case among other songbirds. But instead of hunting insects for her, he was out hunting and bringing back waterfowl feathers, apparently from bogs or shorelines that lay miles from the clearing. These places are also where the swallows' food comes from. It seemed as though feathers might be delivered into the nest as a substitute for food — perhaps feeding was the evolutionary root of the behavior. Instead of bringing a dragonfly or a mayfly to the female that was incubating the eggs, the male supplied a substitute. Conditions in places like a beaver bog may have induced this trend. Presumably many insects would be close by, the female could easily get them herself, and so the male's role in feeding became less critical. His drive to collect insects could be redirected to something else.

12

DISAPPEARING CHICKS — 2016

THE FIRST EGG HATCHED ON THE MORNING OF JUNE 6. To my great surprise, it was the chickadee egg, the one I thought might be rejected and shoved aside because it differed hugely from the swallows'. Unfortunately, the chickadee nest had in the meantime failed (due to an invasion of red ants), and so I left the baby with the swallows.

The next morning at 6:50 a.m., the male voiced isolated *chilk*s as he circled high in a clear sky over the windless clearing. The female was sitting tight on the nest, and as I photographed her there she did not budge. Finally, as she peeked out from the nest-box entrance, the male was no longer near. She went back down but came up a minute later, and then again went down. Maybe she was expecting him back and getting anxious, looking for him. He did not return until thirteen minutes later, giving a single *chilk* while diving down out of the sky; she flew out the moment before he zipped in. A minute or two later he came back out, and I checked the box. To my great surprise — the just-hatched chickadee chick was missing! One or the other of the swallows had either swallowed it (no pun intended) or carried it out, apparently within hours of its hatching. The female swallow may not have been able to *not* incubate the chickadee egg, even though it was on the periphery of the nest. But she could dispose of the fledgling.

After a night of thunder, lightning, and torrential downpours, June 8 dawned overcast, windy, and cold. Trees whipped around, branches slashing every which way. Observations became difficult to make under the thickly overcast sky and amid the noise. I was limited to doing just one thing: keeping my eyes glued on the box.

At 7:36 a.m. a swallow flew out, and I ran to check the nest; still five swallow eggs and one phoebe egg. At 7:45 a.m. the female zipped back into the nest. This — leaving on her own — was unusual; she always waited until the male was back before departing. She had probably left because she was hungry; he had not shown up to relieve her to hunt for her food. She waited until 8:46 a.m. before making a few cheeps, and then she again came out alone. This time she landed on the locust tree's high perch, but he was not there. She had not perched there for many days, as she preferred the low, near-to-the-nest solar panel. After perching in silence for twelve minutes, she flew around the clearing and then disappeared in a southeast direction. In six minutes she was back in the nest-box, but at 9:18 a.m. she left again, returning in just one minute, and then she aggressively dive-bombed me. She had for a long time ignored me. Was she frustrated with the male swallow's absence and taking it out on me?

She left a half-hour later, this time apparently with a distinct destination; she flew in silence in a direct line west, without prior circling. Back in fifteen minutes, she went straight down into the box. The sun was by then peeking out of the clouds. Still no mate. A half-hour later, at 10:34 a.m., she again left in silence, heading directly west, and returned in a half-hour — a long time to be off the eggs. This behavior of the female tree swallow undoubtedly accounts for the variable duration of incubation that has been reported for this species.

Expecting not much of anything new, I allowed myself a break and decided to do only spot checks on a predetermined schedule. As per scientific protocol, I'd check the box every half-hour, on the half-hour. The notes that resulted were boring: at 1, 1:30, 2, 2:30 p.m. — no bird in or near the nest-box; unfortunately I had no idea of what happened between these checks. But at 2:45 p.m., alerted by a couple of *chilk*s, I saw three swallows flying around, and soon just two, as the resident female entered the box and stayed to incubate. Several minutes later the male returned to his usual perch on the black locust tree.

After this, the female's departures and returns were usually separated by only a few minutes. The red dot on her bill was still clearly visible, though her chest paint had faded or been preened out. Finally, at 3:21 p.m., when she left, the male dived into the nest-box, but after only a minute he came up to perch in the entrance — for a total of fourteen minutes. He acted odd: at times he tucked himself deep into the entrance, then leaned far out and looked all around. Finally, he *chilk*ed loudly, signaling that he was about to leave, despite the fact that the female was not yet back. And so it was: he leaned out ever farther and then left. He had been waiting for her, but his patience ran out.

On June 9, a strong north wind blew all day, and temperatures stayed in the 40s. The male swallow perched briefly on the nest-box at 7:45 a.m., then left, and she entered. In all of my fifteen checks into the box on this day, I found the eggs unattended. The female may not have appeared until 11:40 a.m. and showed up again at 6:20 p.m., and he came twice also. Once when she was absent and he perched on the panel, I tossed out a long black feather. He ignored it. I then tossed a white one, and he nabbed it as it left my hand. The male flew to the box entrance and dropped it in; in the same motion he turned and flew fast straight into the distance. He had someplace to get to, in a hurry. That he had stopped to drop the feather off at the nest made my day. I tried the same test the next morning, but he then showed no interest.

On June 10 the wind finally stopped and the sun came out. And this warmth was lucky for the about-to-be-hungry chicks; the first three babies hatched within about two and a quarter hours (the first at 10:40 a.m., the second by 11:14 a.m., and a third by 12:55 p.m.). Both swallows carried eggshells out after each hatching. There seemed to be a division of labor in their foraging now: she flew mainly low and around within sight of the clearing, while he after each nest visit flew off toward the southeast.

On June 11 the last baby hatched at 7:10 a.m., eighteen and a quarter hours later than all the rest. The chicks hatch by chiseling a line to weaken the shell from the inside; then they exert pressure to crack the shell into halves. The male swallow left the nest-box, carrying half an eggshell in his bill, and he dropped it within several meters of the nest. That other half of the eggshell was still in the nest, but immediately after he removed the first, the female flew in and came out with the second. He then perched on the panel, and she went back into the nest-box and stayed there, not even bothering to chase off a chickadee that came and peeked into the box.

It was likely one of the best days in a long time for the swallows to find insects in the air, and I wrote an enthusiastic "Yea!" in my notes, even though I had been a sitting victim for the black-fly scourge while keeping track of each entrance and exit. One or the other swallow was in the nest-box at all times. They were now busily feeding the young, which were raising their tiny heads, opening their pink mouths, and chirping faintly, even when I opened the box for a peek.

The next day (June 12) could scarcely bring anything of more interest. With the young hatched, it was long past the time for nuptial gifts, copulations, or possible nest parasitism. But in the garden I put out four white feathers and four black. At 6:12 a.m., when the female left the nest-box, she made a low pass over them, then picked up a white one and returned directly to the box with it. Over the next hour another white feather was taken also, while the four black remained on the ground.

Rain had pummeled our roof that night, and it was cool and overcast on June 12, a no-fly day for insects; half the time there was no swallow in the nest. Foraging now demanded much of the parents' time; it took precedence over staying in and brooding the young. Whenever I opened the nest-box to check on them, the babies were wildly peeping and craning their necks, their eyes still skin-covered and unseeing.

On June 14 the northwest wind was cold and unrelenting, and in the morning the babies were piled into a lifeless-looking heap. The parents were presumably foraging; neither one could stay to brood the babies when food was a priority. By 11:30 a.m., with the same weather continuing, the young stayed huddled together. There was still no brooding, nor any interest in the new feathers, white or black, that I had laid out. I thought the parent birds might now want them for nest insulation.

The next day was gorgeously sunny and warm. I watched for an hour, from 10 to 11 a.m. This time the pair of swallows made a total of fourteen trips to the nest-box. They often lingered in the entrance, and their trips away lasted only two to five minutes each. Food must have been close.

The weather continued to be beautiful over the next five days, and the young had grown to nearly full size and weight. But they were still pink and lacked feathers. The parents seemed to have time to spare, sometimes circling high in the sky above the clearing. I estimated that fledging would take place in the first week of July, when I was committed to spend a week on Star Island, one of the Isles of Shoals, off the coast of Portsmouth, New Hampshire, but I planned to be back in time to see the fledging.

As soon as we got back home to our clearing, on June 26, I went to see the swallow family that I had watched so closely for two months. Immediately I felt something was wrong. A young swallow should have been perched in the nest-box entrance, anticipating a meal. It should have been chirping, at least occasionally, as a parent flew by. But no swallow at all was flying. All was still, though the babies could not have fledged — that would not happen for a week or so. I anticipated disaster even before looking into the box, but was still shocked to find that the nest was disheveled and contained no swallows. Neither did it contain their droppings, which normally accumulate the few days before fledging.

The nest had been robbed. Looking for clues, I remembered the scenario of two years earlier, when I had at first attributed the killing of three ready-to-fledge young at this nest-box to a large black carrion beetle, later realizing that a raccoon, reaching its claws into the nest, was the more likely culprit. I therefore immediately checked the nest-box post for claw marks. I found none.

Red squirrels and flying squirrels are major predators of eggs and nestlings in these woods, and both species had been coming in numbers to our bird feeders this year, to partake of a generous and constant supply of black sunflower seed intended for birds. That seed supply would not have been replenished in our absence, and the regular partakers of the mammalian sort would have searched for other fare. Squirrels of both types are small enough to enter nest-boxes. Either could have taken out one nestling at a time and cached some for future meals; I had watched red squirrels do this routinely when they were harvesting chestnuts. I had seen a flying squirrel raiding a sapsucker nest with young.

Clearings, inundated as they generally are by the water of a beaver dam, are not natural habitats for red squirrels. This may be part of the reason why swallows have evolved to use such habitats. Squirrels would not be a problem in the water-killed trees in the flooded area. Tree swallows nested in a beaver bog about two kilometers from the clearing, where tall old naked tree trunks stand; they have cavities made by three species of woodpeckers. The swallows never nest in tree cavities within the woods surrounding our cabin because they would be routinely victimized. But this year, in this clearing, both the swallows and the squirrels had likely been lured out of the woods.

With this sad outcome in mind, I immediately wrapped a metal flange around the nest-box pole, to try to prevent such nest predations from happening again. It would of course be almost a year before the swallows would be back for another try at nesting, if in fact they did return.

13

DOMINATING OF THE AIRSPACE — 2017

So far in this study, no more than one pair of swallows at a time had nested in the clearing, though many individuals and pairs visited it. These visitors may have been previous territory holders, offspring from an earlier clutch, former mates or associates, or naive wanderers of either gender, seeking mates or potential nesting sites. The resident birds recognized individuals and treated them probably according to their relationships with them or the visiting birds' apparent intentions. But who stayed, who left, and why? I had some hints about the dynamics in play, though the outcomes were not always predictable. The result, however, was always the same: just one pair remained to nest, making every spring a jousting tournament on the wing.

The snow in the winter of 2017 was deeper than most I could remember. Winter dragged on through March and into early April, as morning temperatures stayed below freezing and the snow that softened on some days would freeze to a thick crust at night. But spring weather arrived suddenly on April 10; by noon, air temperatures had shot up to 65 degrees F. The bees from our hive had their first mass exit flight, and a tom turkey gobbled. A robin sang at dawn, and our phoebe was back, fluttering at the old nest site under the eaves of the log cabin, where a pair nests every year. The swallows are never far behind, if not ahead, and one sounded calls of *chilp, chilp, chilp* as it circled around the cabin and the clearing. Barely an hour later, this same swallow or another landed to perch in the entrance of the last-used nest-box, which now had a metal flange around its pole. The bird then ducked its head and upper body in and out at least twenty times, as if wanting to enter but not daring to go all the way. It was in shiny adult plumage — not a young swallow from the past year. Was it one of the pair from last year's failed nesting, remembering the disaster? Or perhaps it was one of the associates that had been tolerated after incubation had started? I had no way to determine which one it was, but previous observations suggested possibilities.

Cold air and overcast skies prevailed during the next two weeks, and visits from the swallows were rare. But then a male returned, with a female in brownish garb. He hung at the entrance of the same familiar nest-box, peeked in, and sang his gurgling triple-note song. The female joined him there to perch on the box as he flew back to the solar panel to sing.

After many minutes he stopped singing, and she flew to the panel and then left it to fly to a previously unused nest-box, one on the other side of the panel. He then immediately stopped singing, and when she returned, he fluttered over and at her, in a seemingly aggressive way that escalated into a violent chase over the field and through the woods, before both finally flew to the locust tree. They perched at opposite ends of the branch, and there was no more singing. From her feather garb, I could tell that this was a young female. The male had been trying to show her a nest-box he found suitable, and she had initially shown interest, but then examined another box. He could have been agitated by her rejection of what, to him, seemed a suitable nest site, based on either previous experience or recent viewing.

Then, over three days of rain and heavily overcast skies, the swallows were entirely absent. Finally, on May 4, two did come, but their apparent indecision gave the impression that they were a different pair, or at least one of them was new to the clearing. They went back and forth between two boxes at intervals throughout the afternoon, until the male finally entered a third, and then a fourth box. This female was glossier than the one I had seen three days before. Neither had so far landed on the closest tree, the previously much-used locust, or the solar panel — the almost exclusive perches of the past year. This behavior suggested that both swallows were newcomers here.

ON MANY DAYS of early May, the swallows were mostly absent, except for a few drifters, who could not stay because of the bad weather. Nothing of note occurred until May 12, when a new pair arrived. One perched where no swallow had ever perched here before: the tiptop of a huge maple tree in the clearing. The female of this new pair had a noticeably brown head and backside, and she was much easier to distinguish from the male than were most females. I hoped no bad weather would chase this couple away because her markings would greatly improve my ability to tell the two apart at a great distance. As I had hoped, the pair returned repeatedly and always perched together on top of the maple. I would see much of them in the two months to come. But this did not guarantee that they would nest here — one or another of the recently observed pairs might return for another try at securing a nest-box here, or a new pair that I had never seen before might arrive on the scene to test their luck in the same endeavor.

Already, over the next two days the most recent pair often moved between the chosen nest-box and the perch on top of the maple. Then, on May 16, a day of sun after previous torrential rains, I heard no swallows when the sun was rising over the ridge. But an hour later, loud *chee, chee, chee* alarm calls erupted, and two, then three, and then four swallows swirled around, in a huge fight. Two seemed to collide in midair, tangling in a twirling ball that spun to the ground. The antagonists remained engaged even then, for about a half-minute. Both pairs had perceived that they owned a hugely valuable asset — this clearing, with a nest-box — and therefore needed to protect it. The fights occurred four times in a row, after which only one pair remained. It was the maple-tree couple. The male perched there and sang, and as before, he repeatedly dived down to perch in the same nest-box entrance. He then looked up to the sky several times, flew up again, and then returned to the box entrance, while the female chased off what looked to me like another female.

Not much changed the next day (May 17). Only this one pair came to the clearing, and the female spent much time soaring while the male perched at the top of the maple; the male repeatedly alternated between two perches: the maple tree and the nest-box entrance. He flew up to fly with the female, then resumed his nest-box perching. Or he sang from the maple in a faint low whisper. The female didn't appear to respond, remaining silent. However, at 11:15 a.m. she landed on the bare ground in our garden, picked up a tiny piece of grass, and flew into the box. He then sang his whisper-song from his high maple perch, and when she left the box, he dived down and passed low over the garden, and then they left together.

They returned near daybreak the next day (May 18) but did not stay long. It was a lovely day, at 80 degrees F, with hazy sun. The female swallow landed on the garden soil and picked at the ground there, without retrieving anything. Then she flew around in the clearing before disappearing out of sight. The maple tree had by now leafed out, and the male switched his roosting spot to the still-bare black locust tree. The swallows prefer perches with a view, and without leaves. He sang there softly as the female again landed in the garden, then flew into the box, without carrying anything in her bill. For a half-hour she repeated this behavior, only occasionally carrying in a tiny fragment of a rotting blade of grass.

On May 20 the male swallow fluttered over the female's back in what looked like a mating attempt, but she did not crouch and raise her tail, the typical ready-to-mate position. Instead, a highly animated and vigorous chase ensued — the kind I had seen the two previous pairs engage in at the beginning of mating. The sheer vigor of the pursuit looked aggressive, but there were, as before, no vocalizations — not one *chee, chee* call, like the ones that characterized an aggressive pursuit of newcomers. The birds were silent. As I had in previous years during these mating-time chases, I could not see a third party anywhere. It indeed appeared that only the two were involved. The first two times I had seen such an energetic pursuit through the skies, I simply could not believe my eyes. Was the chase something like a test of agility in flight, which could make a life-or-death difference in the ability to catch insects and feed future offspring? The two swallows were, however, already a couple. So it did not fit: the purpose of this chasing remained a mystery. Perhaps this behavior marked a commitment to nest building.

Immediately after the wild chase, the male went to the nest-box, where he sang as the female circled and then came down to land at the box. He had entered it by then and now came out and flew to perch on the locust tree, to sing from there. The female went back into the box and peeked out, staying silently in place for ten minutes. Then at 7:03 a.m. she left the box, he instantly departed from his perch on the locust to enter the nest-box, and she flew to the ground and back to the box, but without carrying anything in. She stayed in the box as the male returned to the maple tree and continued singing from there. Finally, at 7:35 a.m. she picked up a tiny bit of rotted grass and carried it into the nest. After this, the male continued to sing at the tip-top of the maple, and the female continued making trip after trip, empty-billed, to and from the nest-box. He then flew to the ground twice and picked at grass but did not pick it up, but still flew to the nest-box entrance to hang there and look back at her. This behavior was reminiscent of the way that previous males had prodded a mate. It seemed to work: in another fifteen minutes, when she was flying to the ground in the garden, she started tugging at grass and made several more flights to the box, carrying pieces into it. The male immediately resumed his perch, and this time became silent.

At 9:10 a.m. the female made several more trips to carry grass. After one, she remained in the box a minute, after which she leaned out, as the male made three swoops close over the ground in the garden. She then came out and again went to the garden to pick at grass. Then she flew high, and he entered the box and sang from inside it, to perhaps make his message even more clear as to what was required to form the nest. But of course, like all the other males, he himself would not engage in the physical building of the nest.

The next morning (May 21), while perched on the black locust, the male fluttered several times over the female's back and touched down twice, and they mated. Then both flew up immediately, soared around the clearing, and then returned to the perch to mate four more times. They then swooped around the garden, and the female landed there, didn't pick up anything, but still flew into the box. She stayed inside several minutes, and then peeked out as he landed in the top of the locust. He sang, and she picked up her first long piece of grass and flew off with it. He followed her as she trailed it in the air before entering the box with it. She then made nine grass-collecting trips between the garden and the box. There was little interruption in this activity, except when I heard *cheet, cheet, cheet* alarm calls as a third swallow arrived. As before, the female sounded the alarm and chased the intruder. And, as usual, the male flew down to guard the nest entrance. The two were absent that afternoon, as well as most of the next day, which was heavily overcast.

On May 24, I placed five long-plumed white feathers on the bare ground in the garden at 9 a.m., and I checked back at 4 p.m.; four were in the nest and one remained on the ground. This nest-box had a floor much larger than the others, and the feathers had not been incorporated into a nest bowl. Instead, they were apparently stored in the space in front of it — a clue that the feathers may have a function in the lining of the nest, not its overall structure.

This new pair had achieved residence in the clearing, but they were still stalling. During one of her stints in the sky, the female spent an hour and two minutes fluttering around nonstop. She may have flown fifty air miles, at least, but she was staying at the clearing, essentially going nowhere. The two had chosen a nest site, but they were not making progress in nest building. Meanwhile, the chickadees were incubating their eggs, and the robins' and phoebes' babies had hatched. Then rain pummeled the roof of our cabin all night long, and the wind and the cold (45ºF) continued for two days. The swallows were absent.

The dawn of May 27 was partially overcast, but the wind had stopped, and at 5:55 a.m. the pair was back, singing at the top of the maple. It became a gorgeous day for swallows (and black flies), and the first thing I did was to put out ten feathers in the garden. The female flew there, landed beside them, ignored them, and picked up a tiny bit of grass instead. She continued carrying more bits of grass into the nest-box — eight in ten minutes, a sharply increased pace. But the nest was still a barely noticeable ring of grassy bits, with several feathers pushed to one side.

But my doubts about whether the pair would commit to nesting vanished the next day (May 28). It was cool and still, with a pale sun shining through a foggy haze at 6:20 a.m. The pair perched quietly on the partially leafed locust tree. For the next ten minutes the female did not move from the spot, but the male repeatedly flew up from his perch to hover over her back, fluttering back and forth, occasionally touching down on her back while continuing to hover. Each mating took about a second.

And as he got into the rhythm of it, he again fluttered back and forth, and again touched down. All this was accompanied by an excited chittering. Six of these hovering sessions took place within six minutes, with mating contacts achieved in each of six, four, six, four, three, and four. This was followed up by three matings. After this, the pair left the locust tree, and both circled and took turns diving into the box, then circled silently together. Then both made visits to the box and resumed mating, first six, then five more times.

A half-hour later, at 7 a.m., the female carried a piece of grass into the nest-box while the male sang from the maple. She stayed inside the box for fourteen minutes. She later brought in five bits of grass but still ignored the feathers. Thirteen feathers were still inside the nest-box (though not in the nest itself), and I now removed them all, in order to find out whether the birds monitored what was in the nest or nest-box on a schedule in relation to the nesting cycle.

It was so cool (45ºF) and foggy on May 30 that I wondered if the swallows would show up at all. But now, having committed to nesting, they could hardly not present themselves, and at 6:10 a.m. they did. I heard the first *chilp* and saw the pair in the air.

Within a minute the male entered the box, staying down nine minutes before peeking out. This was surely enough time for him to assess the nest contents, if he needed to, in order to go hunt feathers. After he came out, he spent twenty-six minutes idling around the clearing, and then he slipped into the box once more, staying down for six minutes. His flight had been leisurely and silent, with periods of soaring on fixed wings between spurts of fluttering. No other birds were contesting the clearing — this duo now obviously owned it. He *chilk*ed and chattered as he flew by the box, so he (and I) knew the female was in the box; otherwise he would not have circled over the box for so long, nor announced his presence when he flew by it. However, she did not come up to the nest entrance, which meant that something had changed. At 7:19 a.m. she finally peeked out, then left, in fast, straight flight, into the distance to the east, and he flew after her. The nest was empty — still no egg. An hour later both were back, and she made repeated trips to the garden. But this time she landed at a different part of it, next to the blueberries mulched with pine needles, and now she collected these pine needles for the nest. Each needle was considerably longer than any of the bits of grass she had carried in so far. She now used this material exclusively to make the entire nest. For now, the feather lining was still in the future. But might the swallows show some interest in them now?

14

FEATHER PLAY — 2017

TREE SWALLOWS ARE WELL KNOWN FOR DROPPING FEATHERS in the air, catching them, and releasing them again, and, when other swallows are near, taking them from each other. The meaning of this behavior is unclear, and so it is allocated to the category of play, and perhaps thought trivial. As I offered feathers to the swallows, I inadvertently created a record of when and where and under what conditions these behaviors occurred and how they transpired, simply because I wrote everything down. There is no knowing beforehand which behaviors might have meaning, so I recorded them all, always assuming that everything and anything could be relevant. Would "feather play" happen this year, as it had in the past, after the nest looked finished? I set five white feathers onto the ground in the garden and watched. It was near noon, May 30.

The male swallow dove off his perch on the locust tree, picked up one of the feathers, and flew to the box entrance with it. The female was circling overhead but didn't respond. Instead of leaving the feather in the nest-box, however, he flew up into the sky with it and circled. He then came back down and made repeated landings at the nest-box entrance. Again he flew off and circled while still carrying the feather. Still she made no response. He then repeated the behavior two more times, never letting go of his feather and singing intermittently. But on his fifth trip to the box, he was no longer holding the feather; he had dropped it somewhere. The female then flew down, landed in the garden, picked up a pine needle, and dived into the box with it. She did this twice. Then both left. There was so far still only a small ring of reddish-brown pine needles in the box — only the bare beginnings of a nest. As her drab coloring indicated, she likely had hatched the past year. She had never built a nest before, and it looked as if she didn't quite know what she was doing. Or perhaps her hormones had not yet triggered the love for feathers.

On the next day (June 1), however, my notes during two hours of watching spanned four densely handwritten pages. The day had started with a clear blue sky, with wispy clouds drifting over from the west. It was wind-still and a comfortable 50 degrees F, and the black flies descended on me in greater numbers than I believe I had ever experienced. My hands were smarting from bites as I took notes. All the while my head needed to remain uncovered, so I could see all the details of the swallows' behavior. It was now crucial to distinguish the male from the female, and to keep track of both of them continuously and at the same time, in order to follow the story I wanted most to follow. Soon enough we'd be into the long and boring routine of egg laying, incubation, and rearing of the young.

The pair arrived at 5:40 a.m., and the female almost immediately entered the box, as the male continued circling and chirp-ing constantly. He kept this up steadily for forty minutes, when she finally left the box and landed on the locust tree. He then immediately descended, fluttered over her back, and touched down to mate six times. Within five minutes she returned to the box, and he entered right behind her, but then he returned to the tree. She then left the box and went back to the tree also, and they mated four more times before she returned to the box.

Fifteen minutes later he flew down and, in passing the box, gave his *chilk* call, then flew back up into the sky and circled there in silence for ten minutes. Then again he dived down by the box, and *chilk*ed to his mate, and returned to the sky to circle some more. What might he be communicating? That he is still around and has not left her, despite the possibility that he soon may be absent for long durations to hunt for feathers? I doubted it, but it made me wonder if he might at last have interest in bringing her feathers. So I tossed five feathers onto the ground in the garden. He instantly dived for them and, chittering, ex-citedly grabbed one, flew straight to the nest-box, and carried it in to the female. I heard soft chirps within the box then, and he stayed inside for seven minutes before coming back out.

Within another three minutes he took a second feather to her, and after that immediately brought her a third. He then perched in the locust tree and sang. He ignored the last two feathers, which were still plainly in view. Did he want a different one? I put another one down. It too was ignored.

The female had in the meantime left the box to soar over the clearing, ignoring all three of the feathers that were left on the ground. Instead, she took a pine needle. Right after this, she re-turned to the sky, and the male swooped down and took another feather. But this time, instead of flying into the box with it, he began to circle with it. Higher and higher he took the feather, and then he dropped it. It started drifting down, but this time *she* picked it out of the air, swooped down, and took it into the nest. He then picked up one of the two remaining feathers, and again he circled high into the air with it. It seemed like minutes passed as he kept circling with the feather before he dropped it, caught it as it drifted down, and released it again. I started counting: he repeated this maneuver eleven times before she finally appeared, out of nowhere, and grabbed it out of the air. I heard his song then, as she carried the feather into the nest-box. Five minutes later, she still had not come back out. But he had by then picked up the sixth feather and brought it to her by flying directly into the box with it, this time with no catch and release at all. After that, she resumed collecting pine needles on her own. When I then checked the nest-box, the feathers had not yet been incorporated into the nest. They had been left in the entrance of this ample nesting space.

Dropping the feather eleven times could not signal mere clumsiness. It seemed deliberate. The male had brought his mate something relatively hard to get, perhaps like a groom providing his bride with a bouquet. And he had made a show of it. His performance might have symbolized his motivation or ability to provide something of value. If he could chase after and provide white feathers, then surely he would be willing and able to provide mayflies and possibly even high-flying dragonflies for her and the soon-to-come babies. Simply bringing in grass could not make such a strong impression of gumption, interest, ability, or commitment. As with most hypotheses, however, I needed to consider possible objections. If the feather display was this type of demonstration, then why was it timed to occur now, *after the pair had already committed* to nesting and had mated? The feather display should have taken place *before* any nest building had begun, and certainly before mating. This objection then made the behavior even more intriguing, as it suggested the scarcely imagined.

The pair came at 5:15 a.m. the next day, June 2. The female flew into the box and stayed in, as the male soared and remained silent except during the four times he flew by the box, when he made a faint chirp. However, she stayed down in the nest. Not until almost two hours later, at 7:10 a.m., did he finally come down from the sky to the nest-box entrance. He *chilk*ed there once, slipped in, came right back out, and then flew directly to the locust tree. Now she finally peeked out from the entrance.

I then pulled two feathers out of my pocket and dropped them onto the ground. The male instantly responded with excited calls of *cheep, cheep, cheep*, the kind used during chases, and he swooped down and over the feathers six times, all the while cheeping. She then left the box and circled in the sky, and he landed by a feather, picked it up, started to circle with it, and dropped it. She caught it and dropped it also. This time in their ensuing swirl I lost track of who was who, but the feather was dropped and caught ten times before she flew with it into the box. Both then left, and I took the opportunity to open the box and look in: the first egg, shiny bright white, lay in the nest on the long red-brown white-pine needles from the garden mulch. The feather she had brought in yesterday was still in the same place, at the periphery of the nest.

15

EGG LAYING AND
INCUBATION — 2017

THE FEMALE WOULD NOW CONTINUE TO LAY AN EGG A DAY until completing the full clutch of five to six eggs. If the nest was parasitized, I would be able to detect it only if more than one egg appeared in the nest on the same day — as long as the female who dropped one in didn't take another one out. Traffic in and out of the nest needed to be recorded, along with the birds' other activities.

EXCITED *CHEET, CHEET, CHEET* alarm calls sounded barely five minutes after more feather play in the sky on June 2, at 7:35 a.m. The calls indicated a chase, and I saw the male and three other swallows. The male approached the nest-box entrance, perched there as if to guard it, but then flew up to the other swallows, and they mingled in the air. When I checked the box, the female was inside and then flew out, but there was still just one egg in the nest. She flew two loops and went right back in, and he entered right behind her. I then heard soft churring cheeps. Within a half-minute he flew back out, circled high, and disappeared to the west. She stayed in the nest.

From the precedent set by earlier pairs of swallows, I expected their morning ceremony and greetings the next day (June 3), and got up and out at 5 a.m. to see them. It was a chilly 42 degrees F, under a heavy overcast sky, with a westerly wind and thick drifting clouds. At 5:10 a.m., hearing a few faint cheeps from up high, I searched the dark sky and located, presumably, the resident male flying high and fast (the female was in the nest, laying the second egg), erratically fluttering in and out of the clouds. The flight seemed frenetic, and so it continued on and on, in the pattern I had now come to expect from previous nestings. I kept my eye on him continuously. In and out of clouds, hundreds of meters high, he continued nonstop for the next forty minutes. Then at 5:50 a.m. I heard excited *cheet, cheet, cheet* calls — and for a few moments saw two swallows. I immediately thought, "Darn, I've missed her coming out of the box to meet him." But then he came out of the sky, dived into the box, and came back up to perch in the nest entrance for the next twenty-five minutes. I knew she was inside because he at times ducked down, and I heard their soft cheeping and occasionally also his singing — proof she was there. He had not chased the third swallow. He didn't need to; his mate had been present in the box, to keep any intruder out.

It was cold. No insect was flying, and none would have been in the swirling clouds where the male swallow had fluttered. The sky was, however, now clearing, and when I left briefly and returned once more at 6:40 a.m., the male was still fluttering above the clearing, as he had been most of the morning so far, and the female was still sitting tight on the nest. As expected, when she left the nest a little later, it contained the second egg. There were so far no feathers around or under them.

The female did not reappear until late afternoon and again overnighted in the nest, with her two eggs on the pine needles. The male arrived in the evening, peeked into the nest, and then left in a westerly direction.

I expected the male's feather-gathering fetish to kick in now, and just to be sure, I put ten white feathers on the usual place in the garden. The female was in the nest, and the male was circling, but before the test could begin, he was interrupted: a merlin shot across the clearing. Both swallows chased after it, and I heard their *cheet, cheet, cheet* alarm calls receding into the distance. The swallows had quickly distinguished this small falcon from a mourning dove, which the swallows routinely ignored — another tribute to their keen ability to make distinctions among different birds. Merlins and mourning doves are of the same size and have pointed wings, a long tail, and similar muted colors. How did the swallows know one from the other?

The pair returned in minutes from chasing the merlin. The female entered the box as the male landed on it and sang briefly before also going in. Nine minutes later he finally peeked out for a couple of minutes, then flew high, and returned to resume carrying in feathers. Now he never once dropped a feather but flew directly into the nest-box with each, where the female was laying her third egg.

Both soared later that morning, no longer constantly shifting between soaring and fluttering, their usual flight pattern. They stayed near and occasionally entered the box. I put out more feathers, as the ten I had placed on the ground previously had by then been taken to the nest by the male. He now resumed carrying the feathers, and she flew low over them and picked up one, but then dropped it before flying off. The nest was finally feather-lined when I checked it at 4:10 p.m., after both swallows had left: the eggs were hidden from view by the long plumes, placed so that they curled over the eggs. The fourth egg was laid by 9 a.m. the next day (June 5), and the fifth and final one the day after that (June 6).

By June 7 the clutch was complete, at five eggs. Incubation could begin, but the female still left the box repeatedly throughout the day. The nest thus seemed unguarded, but chances for a successful cuckolding were now diminishing quickly, and I suspected that nest guarding would stop entirely when incubation started. Indeed, there was from then on little activity at or near the nest. There were no chases and no aggression.

Feathers were still accumulating in the nest, and since they were duck and goose feathers, they must have been carried a long distance. On June 11, I removed all forty-seven of them and at 9:15 a.m. the next day, ten days after egg laying started and now about a third of the way through incubation, I pulled out another twenty-six. These were also mostly wood duck feathers. But at 3:40 p.m. on the same day, three new ones had been added, and the next morning (June 13), there were seven more. All the feathers were now being incorporated within the nest bowl.

At 7:55 a.m., when the male arrived holding a large white feather, he *chilk*ed. There was no response from the female in the nest (she was likely absent at the moment). He then landed at the entrance and *chilk*ed again and, hearing no response, flew off, carrying the feather. He was still flying with it three minutes later when she arrived and joined him. He then dropped it, and she picked it up and brought it into the nest-box. (I had seen no release-and-catch feather play in the air since the day the first egg was laid.)

She left the nest at times for up to twelve minutes, but he did not take her place in the nest-box. Instead, he pursued his feather fetish. The next time he arrived this morning (8:51 a.m.) she was gone, and after he flew to the entrance of the nest, he did not drop his feather in but instead left, still carrying it, and circled with it. He left the clearing only after she had returned. Eighteen feathers had been added by June 14, halfway through incubation, for a total now of sixty-five feathers. I again removed them (June 15). But the nest was newly lined with feathers by 5:30 p.m. of the next day (June 16).

The 17th was too cool and overcast for insects to be flying. But the female swallow brought one wood duck feather, leaving behind the five white ones (also duck) that I had dropped onto the ground, directly in front of the nest-box entrance. These feathers did not go unnoticed, though. While perching in the nest entrance when the female was gone, the male repeatedly cocked his head and looked down at them.

Ten new feathers were brought into the nest-box on the twelfth day of incubation (June 19), but these now remained in the nest vestibule, as had been the case at the beginning of feather gathering. It seemed negligent of the female, but it was time now for drastic change, since the eggs could burst open to release their pink babies today, tomorrow, or the next day at the latest. Each baby within its egg would by now have a special tool — the egg tooth at the end of its bill, used to scrape a circular line that weakens the eggshell. Then, under pressure exerted by the baby's neck, the shell gives way, popping into two halves. Then it would be time for the male to help feed the young. Would he then stop bringing feathers?

On June 20, the thirteenth day of incubation, the female sat tight on the eggs at 6:45 a.m., with five feathers in the box vestibule in front of the nest. Nothing happened until sixty-three minutes later, at 7:48 a.m., when, after hearing several *chilk*s in succession, I saw two swallows in the air. The female instantly responded, leaving the nest-box and her incubation to join them. There were no *cheet, cheet, cheet* alarm calls or attempts to give chase. Instead, all three swallows then left, seemingly peaceably, together. Seven minutes later the female returned to resume incubation.

The swallows' behavior had been remarkably consistent with what I had observed before. Patterns were falling out, and that was reassuring.

16

LAST BUT NOT LEAST — 2017

BABY SWALLOWS ALL LOOK AND SOUND ALIKE TO US. WE perceive them through our human senses. But although swallows have, like us, the same senses of vision, hearing, and likely taste, these and other senses are more finely tuned in relation to important aspects of their lives. They likely see, for example, ultraviolet light as a color, and also have sensations to perceive magnetic waves and barometric pressure. Perhaps adult swallows can also differentiate one of their chicks from another. My previous observations suggested that they remember and search for specific individuals. An opportunity to confirm or reject such a possibility might soon come up this nesting season.

THE FIRST BABY hatched between 6:30 and 7:08 a.m. on June 21. Feathers were still in the nest vestibule, not yet incorporated into the nest. A half-hour later, a chickadee flew to the nest-box and peeked in — the swallows ignored it. This bird was no longer a threat, just like other swallows, which were not anymore competing for the nest-box. Again I saw three swallows flying. One approached the nest-box entrance, fluttering by several times in succession, but was too hesitant to enter. Why was it still taking an interest in another swallow's nest?

Three babies hatched by noon, and two eggs remained unhatched. The fourth hatched at 3:50 p.m. By late afternoon the four babies were holding their heads high and gaping vigorously, their bills open, their skin-covered eyes unseeing. They were responding to a disturbance — my opening of the box. I was surprised that one egg had not hatched. All are supposed to hatch more or less synchronously because development of the whole clutch starts only when incubation kicks in — normally, only after all the eggs have been laid. Typically, they are all on the same schedule. Wouldn't the late bloomer of a clutch suffer a huge disadvantage?

I got up, as on the day before, at first light on June 22, the longest day of the year; it was full daylight by 4:55 a.m. The female was bright-eyed; she hunkered down on the nest, but the male did not come until 5:27 a.m. As usual, he announced his arrival in the clearing with a single soft *chilk* and continued circling on high. Then, after four swoops, he dived down into the box, and she then exited. He stayed inside for two minutes, and when he left, I checked the box: the last egg was still unhatched.

The next day (June 23) rain was pouring, and the female did not leave the nest until 7:05 a.m., flying immediately in her usual southerly direction toward a beaver bog, the likely source of the wood duck feathers. The male had been in the box with her as well, and he peeked out from it the minute she left. But he did not leave until twelve minutes later. Then, in four minutes he returned, and within seconds he peeked out from the entrance, making twitchy head movements and looking all around. He left sixteen minutes later when the female returned from her trip, which had lasted thirty-three minutes, her longest time away since before incubation had started.

THIS WAS THE LAST I saw of the swallows for nine days, due to another long-scheduled trip (this time to Poland). Unlike the many types of songbirds that leave the nest when they can barely fly, the young of tree swallows take a longer time to fledge. Swallows (and swifts) must be able to fly well enough to capture flying insects perhaps a day or two after they leave the nest. As had been the case after the trip to Star Island, I should still have at least a week to see what I wanted to see — the transition from life in the nest-box to life in the air.

It was pleasantly warm, with scattered clouds, upon my return on July 2. But I was anxious about the swallows. It had rained a lot; our rain barrels were full. No swallows were around as we entered the clearing, and I feared a repeat of the past year's disaster. But as soon as I touched the nest-box, the slight jarring prompted cheeps from within.

The four chicks looked healthy, already showing some white at the breast; pinfeathers were starting to open at the tips, and the chicks had dark stubble on their backs. They started begging when I opened the box but immediately ducked back down; they had made some kind of discrimination — by means of visual, mechanical, or aural stimuli — that prompted them to respond in this manner. I snapped a picture. And then I thought,

"Wait a minute, aren't there supposed to be *five?*" The number of chicks as such may not make much difference in my documentation of fledging, but it did matter, for an interesting reason.

I checked, and rechecked. There was no doubt — only four. When Lynn and I had left for Poland, the clutch had consisted of four freshly hatched young and one unhatched egg. I expected the fifth egg to hatch later. But there was no sign of the egg or a fifth hatchling. There was no evidence of a disturbance made by a predator; before we left, I had been almost paranoid about that potential problem. The metal flange was still wrapped around the nest-box pole. I had also set twenty mousetraps in the grass near the pole, to startle any squirrel or raccoon that might step there. And just for good measure, I had dumped two bottles of hot pepper sauce all around to discourage a bear.

After eliminating other possibilities, I had to conclude that the parents themselves had likely removed the late-hatched baby, recognizing it as somehow different or as one too many. This resembled the behavior of the earlier pair of swallows that had removed one just-hatched chickadee baby and one swallow baby — the latter not foreign but perhaps different from the others in coloring or size. The keen discrimination and judgment that enable such behavior would have evolved under strong selective pressure related to nest parasitism, and other factors such as lack of food requiring energy economy. The only possible nest parasites that swallows need to watch for are others of their kind, along with cowbirds (no other local birds of similar size pose a threat as nest parasites).

Up at the crack of dawn the next day (July 3), I found partly cloudy skies, still air, and a comfortable temperature of 63 degrees F. I already heard the babies' begging as I approached the nest-box. I was not close enough to disturb them, and so, as expected, I found the female inside the box with them. Nothing very exciting happened until 6:21 a.m., when under a clearing sky, with no parent in the box, I heard a *chilk*. As expected, a swallow dived into the box. But as I scanned the scene, I saw two in the air; there were now three swallows here. Furthermore, when the first one came out of the box, the other two went in and out of it also, each in about ten seconds. Four nest visits had occurred in less than two minutes. The third swallow may have been involved in the nesting all along, and it was likely another female; I had sometimes observed two females here.

On the next morning (July 4) the resident pair again had the company of a third bird. The three soared together for many minutes, showing no trace of antagonism, a huge contrast to the vicious fights I had witnessed at the beginning of nesting season. The presence of the third one was no accident; after the pair had visited the young and left, in less than a minute another female arrived and entered the box. Other swallows had routinely been coming near, but not until almost the *end* of the fledging period did they do so conspicuously and without repercussions, even upon entering the others' nest-box. This was a spectacular behavioral change, at this point regular enough to be considered a common if not routine occurrence, a reality worthy of acknowledgment.

The Fourth of July was a perfect day for viewing, and I kept track of the swallows' comings and goings at the nest entrance. As before, I was able to distinguish male from female, but not one female from another. My fourteen pages of condensed notes over the next ten days did not contradict my earlier speculations. Different females visited the nest-box, for any length of time, even up to an hour or more. I couldn't assume that the nest's resident female would necessarily make more foraging trips than her mate to get food for the young, but of the 320 trips I monitored, 187 (or 58 percent) were made by a female.

By July 13 the young were completely feathered out. They would fledge after their wings grew slightly longer, on perhaps the next day or so. From previous observations I suspected that the parents would soon induce the young to leave their cozy quarters by reducing their rations and making them hungry. They would then lure them out with food. And for that reason I had persisted with the long tedium of gathering baseline data about feeding schedules, one that accounted for variations of weather, since weather affects the rate of feeding.

Trying to think like a bird (which means being able to predict its behavior), I anticipated that the female would no longer stay overnight with the young. To check, I had dragged myself out of bed at 3:30 a.m. (on July 14), took flashlight in hand, and went to the nest-box. As I opened it and shone the light in, contrary to my expectation the female was inside. She scurried to the back of the box, like a mouse running away. I counted the (four) young, then went back to the cabin to build a fire for an early cup of coffee. The day ahead might be long. Expecting constantly scintillating observations is the best guarantee of tedium. To be successful as a naturalist requires the mindset of a beggar, eager and thankful for every crumb of information.

Throughout the previous week and up until July 13, the swallows' combined foraging trips had averaged 16.3 per hour, but that schedule changed abruptly on July 14. In the morning, it was 14 visits per hour, but by afternoon the number had dropped precipitously, to only 3. One baby then started to peek out from the nest-box entrance, cheeping away, for up to nine minutes at a time. While this was going on, the parents idly perched side by side in the locust tree, silent, preening themselves as though their babies' begging wasn't the least bit interesting. Do the parents "decide" when and how to fledge their young? I had always assumed it was the other way around — that the young go when they are ready and they feel like it.

By daylight the next day, both parents were circling and *chilk-ing*. One baby was again peeking out from the nest-box, but a parent had visited the nest only three times by 9:15 a.m. An hour later both parents were again circling and calling, but the male now shot past the nest-box entrance twice, to then resume his soaring. This behavior was much like his way of inducing his mate to collect grass and build the nest.

The baby's begging at the nest entrance sounded like a cricket's chirping and became ever more emphatic in tone and tenor. I saw the baby's beady black eye glance up at its parents as they unresponsively soared above. It flashed its snow-white breast, a contrast to its slate-gray cheeks and top of its head, and it made quick jerking head motions in different directions. This was perhaps its first view of the world outside the box.

That night when I checked, the four were hunkered down flat in the nest, and they did not move. Fledging would, I felt, happen the next day, July 15. Yet by early that morning, little had changed. Within minutes of noon, though, I suddenly heard a big commotion, as three swallows, two babies and one adult, landed on a dead branch in the black locust tree on the north end of the clearing. Then there was silence, except for the faint cheeping still coming from within the nest-box. The merlin came again, and again both adult swallows chased this agile, small bird-hunting falcon, as they had done during its previous visits. The merlin ignored the swallows, likely realizing it had slight chance of catching an adult swallow that was alert to it, and these swallows were effectively diverting it from the young. A blue jay landed nearby, and both parents dive-bombed it as well, although previously they had ignored all jays.

The two adults and their two young flew up again, and the young not only fluttered but flew almost like adults and then headed for even taller trees, the nearby pines. Then they went back to the locust tree, again accompanied by the two parents. One of our three tamed crows, which the swallows had ignored in the past, came by and landed on an apple tree in the field; one of the adult swallows dive-bombed it. The crow's reaction —ducking as the swallow dived— made it look as if it had been

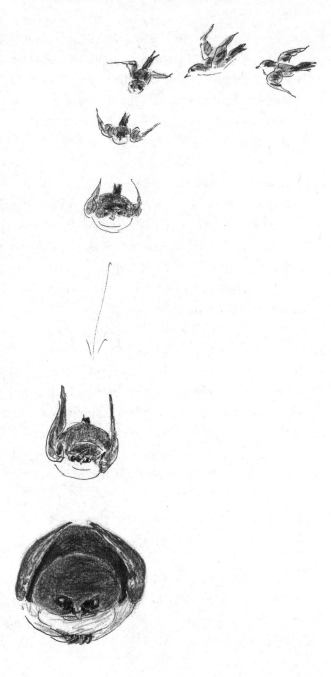

knocked off its perch. To the uninitiated, a swallow flying at your head at forty miles per hour can be intimidating. It is like a twenty-gram ball coming straight at you. When the swallow tucks its wings against its body and drops from a great height, it is tapered like a teardrop, and its feathers are interlocked in a smooth surface that glistens like a polished blue-green mirror. It slips through the air, assisted by gravity, a blue fleck that, in the brief second it shoots at you, allows time for a reflex reaction and nothing more. The swallow comes from behind and produces a staccato chittering noise, changing quickly to a clicking, which you hear only in a fraction of a second, as you turn and duck.

Both parents were now, all of a sudden, regularly visiting the two babies in the tree to feed them at intervals of a half-minute to two minutes. Each time a parent came near, the baby fluttered its wings rapidly and cheeped. The parent fed it in what looked like a flying kiss, as it momentarily fluttered to pass off the food. There were fifty-two such feedings, twenty-six per baby, in one hour. Compared to their rate of entering the box just minutes earlier, this now vastly increased feeding rate was phenomenal. It meant the parent swallows had definitely made an adjustment, a sharp distinction between those outside versus those inside the nest-box; the wildly begging babies still in the box were left unfed.

The babies' cries from the nest-box became louder by the hour. They sounded desperate. By afternoon, one baby was starting to lean out from the nest-box entrance. The male now kept flying by, making liquid calls I had not heard before. The female meanwhile circled the clearing, and the two young that were already out joined her in the air; the three flew off together to the southwest, over the forest.

The male stayed in the clearing and kept on flying past the nest entrance, in front of the constantly cheeping baby that was leaning out. Curiously, the vocalizations from the box sounded to me as if they were made by only one bird. But I had the day before counted four about-to-fledge young. I resisted opening the box for another look because I did not want to interrupt the fledging. Thinking I had missed seeing the other chick fly off, I waited until dark that evening to check the nest, and I was re-assured: both of the two remaining babies were hunkered down there. The two that had left were out of earshot. Would the parents remember the two in the nest? Would they come back here the next morning?

On July 16, I was out early, waiting. No swallow came near until an hour later than usual, giving their typical vocal signa-tures. The young in the nest had before answered an adult's calls instantly, but this morning there was no response from the nest-box. The swallow called, circled only twice, then left. An hour later, at 5:50 a.m., there was a repeat of the same. No response. Twelve minutes later two swallows came. They called, again got no response, and left. One or both of the pair had visited the clearing four more times by 6:35 a.m., and there was still no re-sponse from the babies in the nest. Curiously, the parents had not entered the box to check its contents, as if knowing the nest-lings were not there because they had not responded. But I was amazed at their persistence in returning to the clearing when they had no evidence that their young were still there. Would the parents, hearing no answering calls, at last conclude that the two offspring were gone? Only I, because of my continuous watch, knew the young could not have flown out yesterday. However, given their lack of food, I suspected that the young swallows may have entered into cold torpor at night and perhaps were not yet warmed up enough to respond.

Finally, at 6:47 a.m., the male flew low and close by the nest-box, making liquid-sounding calls. Then he stopped at the entrance, hung there, peeked in, and left. Still no response. The female then arrived and also ducked down in several times, before slipping in. Silence followed, and she came out to perch in the entrance and look out. For a moment I thought I heard a soft churring call from within the nest. But the female left the box, and then three adult swallows circled. Two left, and the third kept circling, leaving only several minutes later.

Only half the babies had fledged. The others were still in the nest-box. What would the parents do next? How would they respond to this dilemma, now presumably knowing that two of their offspring, though silent, remained in the nest? I stayed to watch all day long, taking ten pages of notes as I tried to capture and decipher this ongoing real-life drama. It was worth following all details — an attempt to fathom the mind of the swallow.

At 6:57 a.m. the male perched in the nest-box entrance. This time he held an insect in his bill — I saw its legs sticking out. Both parents were now enticing the two babies to leave the nest. They visited four more times in the next half-hour, always hesitating to enter the mostly silent nest. But eventually, by 7:45 a.m., faint chittering was coming from inside the box, and soon after that, whenever an adult flew by the entrance, I could hear a vocal response. Apparently my hypothesis concerning torpor had been correct; the babies were waking and reviving. Finally, at 8:04 a.m. the male dived in with another insect in his bill and came out without it. He had delivered food! This virtual lack of feeding, except for just one insect, was so far unprecedented. Now the babies' efforts to get food from their parents continued. By 9 a.m. they sounded much louder. Still, neither had shown itself at the nest-box entrance. Did they *still* lack motivation to come out? Or were they too weak? Since these two had not come out with the first two, it was by now nearly certain that they were at least partially starved.

Both parents now appeared to be distressed. They both made the *cheet, cheet, cheet* alarm calls, normally reserved for predators, but they also *chilk*ed when they flew repeatedly by the entrance. Then, at 9:15 a.m., a baby peeked out from the nest-box entrance. The two adults kept circling, but the baby did not fly. It just kept on chirping, and the male finally flew down to it and fed it at the entrance, after what appeared to be a prolonged tease. The female then came down and fed it also. Both adults then left, but the baby stayed there and chirped, and the longer the pair stayed away, the less frequently and more weakly the baby called. The parents' teaser to lure the baby out had not worked. What next?

By 10 a.m. the baby was still in the entrance; it had not gone back down into the nest. The female then fed it four times, the male three times, but not again until a half-hour later, a very long time for a baby swallow not to eat. The parents could surely assess its need for food. But perhaps they also understood the power of deprivation, and the lure of food as an enticement to leave the nest.

The baby eventually went back down into the nest and became silent. But it or the other came back up at 11:55 a.m. to cheep again while the male circled and made *cheet, cheet, cheet* calls, as well as the liquid-sounding calls I had heard at previous fledgings. But this baby didn't budge. It just kept jerking its head and looking up at the male, which then swooped down and flew back and forth directly in front of the box, making *tseet, tseet, tseet* alarm calls. Back and forth, back and forth, but the baby, left unfed, eventually gave up and ducked back down into the box. Both parents then called even more frantically, and the baby came up once more, this time seemingly poised for takeoff — it leaned out. But then again it withdrew, despite the couple's continued flybys at the entrance, and their liquid calls, which by this point took on the urgency of screaming.

It was now noon. The baby needed food. Depriving it of nourishment hadn't gotten it to fly. The adults next tried a different tactic. This time the male landed at the entrance with an insect dangling from his bill, but instead of slipping down to feed it to the young, he just held it there, flew off, and again came back to the entrance, again holding it in plain view. He did this three times in succession. The baby cheeped — and stayed down. Both parents seemed to be flying faster, and the male's *cheep, cheep, cheep* alarm calls were ever quicker and louder. No baby budged. Finally the male hovered at the entrance. In apparent desperation he then fluttered down to the ground, into the grass below the box. Nothing helped. The baby's cheeps seemed weaker, and at 12:18 p.m. both parents left. Then the cheeps stopped.

I have been saying "baby" in the singular because, although yesterday there had been a duet of baby calls, it seemed increasingly clear to me that I was now hearing only one voice from the box. This was, however, an illogical conclusion, since I had peeked into the box the evening before and had clearly seen the two babies. I now peeked in again, and indeed, two baby swallows were present. But one was now dead. An autopsy revealed that it had died of starvation. It had no sign of fat, and its flight muscles were greatly atrophied. It looked like others of the same age that I had found dried up in nest-boxes in the past.

Both of the last two babies of this already reduced (from five to four) clutch had been too weak to fledge yesterday and join their two well-fed, strong siblings, which had flown out of the box and straight up into the treetops. The third remaining live baby of the clutch of five eggs was now balanced on a survival tightrope. It would be close: I could not begin to guess whether it would make it out of the nest-box to fly.

The pair of swallows came back at 12:44 p.m. They repeated the behaviors I had already recorded, except possibly even more vehemently. The male again came three times to the nest entrance, but now he was not holding a small item of food; instead, it was a large dragonfly. Again and again he ducked his head deep into the nest-box, holding this offering in his bill. The single remaining live baby increased the frequency of its calls, from a slow four per fifteen seconds, to fourteen per ten seconds. It was, however, still neither strong nor motivated enough to reach the nest entrance. The male fluttered more and more desperately, dozens of times, at the nest entrance; then it fell down into the grass and fluttered in a very un-swallow-like way. It did no good. The baby stayed down, and the parents left.

In a half-hour the male came back and repeated the same sequence of behaviors. The female had not come with him, presumably occupied with the two young that had flown. After all, she had been the more frequent food provider for them. Again, the baby in the nest did not get fed.

I had by 1:30 p.m. continuously watched the family's struggle since daylight. I had seen more than I could have ever expected of details related to the fledging process, and I did not think there was anything more to be seen or learned. So I got up at 2:30 p.m. — to feed myself.

When I returned an hour later, both adults were calling loudly. They were fluttering not at the nest-box, but over the ground, about fifteen paces from it. There I located the young swallow. It had left the nest-box and apparently flown slightly downslope from it. The parents soon left, and a few minutes later the baby cheeped as the female returned to the clearing. She swooped down and fed the baby as it begged with open bill and quivering wings.

The next morning, July 17, was a calm day, slightly overcast. The temperature was a pleasant 66 degrees F. At 6:30 a.m. a swallow came to the clearing, circled, and called. The baby cheeped from near the end of the clearing, where it had flown and now landed slightly uphill in a tangle of goldenrod. It was still weak, and I picked it up to photograph it. It responded by spreading its wings and snapping at my fingers with its dainty yellow-fringed bill. It seemed a pathetic gesture, but I was moved by its spunk and set it onto a nearby bush from which it might launch. It stayed where I put it, but I later heard it back on the ground among the goldenrod and blueberry bushes.

The parents then came repeatedly, and they faithfully fed the fledgling all day long, never having the least problem with finding it; instantly they knew where it happened to be. When no swallow was circling, the baby chirped, though weakly and sporadically, at three chirps per fifteen seconds. But when a parent was circling and announcing itself with a few soft *chilk*s, it increased its rate of chirps to eight per fifteen seconds, and then to twenty-three per fifteen seconds, and the chirps also became louder the

moment the adult swallow approached. The baby sounded like a cricket, but the swallows were never in doubt as to what they were hearing. They always went directly to the young swallow, which was now finally being fed on a regular schedule. I left it alone but continued to hear it and the parents through the afternoon, as feeding went on. Then, presumably, when the baby had been refueled, with a full stomach and regenerated flight muscle, it flew off in the company of its parents.

BAD WEATHER HAD created a sharp edge between life and death for this clutch, as it had for earlier ones. The parents' likely clutch reduction, from five to four offspring, had taken from the nest the last-to-hatch nestling or an egg, which may have been a parasitic addition to the nest. For the parents, this removal may have meant the difference between the relatively successful outcome — only one of their own had starved — versus losing them all. One thing is certain — the parents cared intensely about the welfare of their young. Swallows are fueled by strong emotions.

I am aware of science's taboo against making assumptions about anything that you cannot prove by measuring. But from the viewpoint of evolutionary history, and therefore comparative biology, emotions are nascent behavior. They activate and direct the behavior that fits an immediate circumstance, without requiring mental calculation, which can act as intervention by override. Emotions are a basic mechanism. We are endowed with many of them at birth. Like other physiological mechanisms, these are shared among species and can be assumed to be present in others, just as we may assume every vertebrate animal has a heart, a brain, lungs, liver, and alimentary canal. We might instead try to see what a swallow sees, and how its specific needs and circumstances are similar to or different from ours. We can also, through empathy, come to understand the feelings of other species, but this empathy must be based on knowledge.

17

THE APPLE TREE PAIR — 2018

SWALLOWS ARE FAMOUSLY PUNCTUAL. THE CLIFF SWAL-
lows at Mission San Juan Capistrano in California, which mi-
grate six thousand miles from Argentina, come back every year
on March 19 (though reputedly "not always"). Friends in Vermont and in Pennsylvania tell me their barn swallows always
show up on May 1. In 2018, my intended last year of focusing on
tree swallows, there was no sign of these birds until May 6; they
were about three weeks late compared to previous years. And
the pair that came then made only a transient flyby. Last year's
local pair, having experienced a possible deliberate clutch reduction, the starvation of one baby, and the threatened survival of
another, perhaps lacked motivation to return there, although it
had not been the site, but rather the timing, that was off. Do they
know the difference? Would they not return at all this year, or return later? Instead, a new pair profited.

• • •

A PAIR OF SWALLOWS arrived very late, at 9:40 a.m. on May 10, and one of them immediately inspected three nest-boxes by hanging at each entrance and looking in. Their reluctance to enter, their examination of several boxes without preference for the one used last year, and the fact that they did not perch on the black locust tree or the maple offered strong evidence that this pair had not nested here before. Then, when both birds repeatedly landed on an apple tree in the clearing (in previous years I had never seen a single swallow go to it), I was nearly certain that this pair had not been here last year.

The male shone a brilliant blue and had a pure white underside. The female could be easily distinguished from him by her contrasting brownish-dun color. While perched (always on the apple tree) he appeared more plump than she. They came six times on this tenth day of May, each time perching in the top of the apple tree, from which the male sallied to hover at and then hang from the entrance of one nest-box after another, occasionally singing briefly. The female showed no interest in any of the potential nest sites that he was visiting. Instead of looking into or entering a box, she eventually left, flying high into the distance, and he then left the latest box he had just visited and chased after her. That was the last I saw of them that day.

I did not see the pair again until three days later, when they arrived shortly after sunrise to perch, as before, on the apple tree. This time, however, the male began to show more enthusiasm than before in visiting and entering boxes, but only after they had spent thirty-three minutes perched next to each other and made one swirl around the clearing. They both left an hour after sunrise, but in twelve minutes they were back on the same spot. Here they vocalized in a whisper, and he commenced that typical liquid gurgling song we expect of tree swallows. He then flew down to enter the box that he had, by process of elimination, started to favor.

This male indicated his positive choice by entering the nest-box, staying inside for a half-hour, singing softly all the while, and then perching in the entrance and peeking out. The female remained silent, perched on the apple tree. Another half-hour later, he finally came out and flew up to perch beside her; they once again flew back to and into the same box. This time he stayed in for twenty minutes, then flew out to perch by her side. But now she left her perch, flew to the box, and hung from the entrance for thirty seconds before slipping in and staying down for eight minutes. The male sang vigorously from the apple tree as she inspected the box.

They then left the clearing together for eight minutes before returning, and he immediately flew to the same box, hung from the entrance, slipped in, and continued singing inside. This time she joined him inside, and he kept up his singing for several more minutes before returning to the apple tree and singing there. She finally came out and landed within a half-meter of him. After perching quietly for fifteen minutes she returned to the same nest-box.

By noon the female swallow had swooped to the ground in the garden, and while making her high-pitched *chilk* calls, she picked up a piece of grass; he sang from the apple tree. She returned to the box and stayed inside for two minutes as he continued his singing: a soft whisper, a barely audible churring. When she then left the box, they departed from the clearing together and did not return on this day. When I checked the box, it contained nothing more than the one piece of dry grass.

On May 14 the male did not arrive until 9:30 a.m. and then perched, silent and alone, on the apple tree. It was not until 11:45 a.m. that I saw the two of them. The female flew by the nest-box several times, finally stopped, and hung from a neighboring box, staying four minutes in that position before flying up, flying high, and leaving the clearing. She again visited the "wrong" box (or possibly the one she preferred?) and then flew *at* the male on the apple tree, as if to nudge him. She did this three times in succession before entering the previously chosen box, the one he had indicated all along. When she entered it, he sang.

On May 15 the swallows arrived at 6:30 a.m. and for the next half-hour reverted to their previous behavior, indicating their mutual attraction to the first-preferred nest-box. At the end of this show of unanimity, the female flew to the garden, landed on the recently harrowed soil, picked up a loose piece of dry grass, and carried it into the nest-box. The male sang. She stayed inside for five minutes before coming up to the entrance to peek around, and there she parked for another twenty-one minutes while he kept up a nearly steady repetition of his chortling liquid song. An hour later, when she finally left the box, she did two more pickups of grass within five minutes, and at 8 a.m. the couple left. The nest-box contained seven pieces of grass.

At 4:45 a.m. on May 16, when it was still dark, the ovenbirds, winter wrens, and the hermit thrush were already singing, but it wasn't until 6:25 a.m., about fifteen minutes after the sun peeked over the ridge and lit up the luminescent fresh greens of the poplars and maples, that the pair arrived. The male perched on the nest-box, and the female took up her usual position on the still totally bare apple tree.

I relaxed on my chair under a red maple tree, enjoying a good view from the edge of the field. A pair of chickadees came by. One landed near my head as I kept looking straight ahead to watch the swallows. It then hopped onto my knee, and from there onto my left arm on the chair's armrest, where the bird picked at a small wound on my forearm. I then slowly lifted my arm to my face, to see what would happen when it looked me in the eye. But that was too much for it — and it left. Nearby in the surrounding forest at least a half-dozen blue jays were making a ruckus. A winter wren sang, and the female swallow made six grass-procuring trips from the garden. The intervals between one trip and the next got shorter. She did her sixth trip in thirty-five seconds, with only five seconds spent in the box. After that she joined the male on the apple tree, where he had been singing nearly continuously all along.

She had picked up only grass stems, each from almost the precise spot where she landed. This was not the pine-needle-toting female of the previous year. Her several circles around the clearing and successive swoops over the garden before landing for a grass pickup were perhaps mere exuberance, or flights to locate specific pieces of grass. Why all this fuss for a piece of dry grass when there were grass pieces almost everywhere? It shouldn't matter where she landed. There were also plenty of pine needles. Might she be looking for something special?

With that thought, I fetched ten white and two black feathers from my cache, and tossed them all at once into the air just as she came flying over the garden. She wheeled, swooped down, picked one out of the air, and took it into the box. On her next pass, with the rest of the feathers now on the ground, she took a grass stem instead. It was within a meter of the stationary feathers. It seemed that, as I had speculated before, feathers in the air were what attracted her. Was this a cue, triggering a redirected insect-hunting instinct? On the subsequent trip she grabbed a feather, then dropped it and resumed her task of picking up and carrying grass to the nest instead. At the end of the day the nest contained about twenty tiny pieces of grass, and one black and six white feathers.

The pair of swallows arrived at about 6:30 a.m. the next day (May 17), and they perched in the apple tree, staying silent for the next half-hour. After visits back and forth between the nestbox and the apple tree, the female rejoined the male on the tree. Later, after successfully carrying in long pieces of grass, both swallows flew high, circling for minutes over the clearing, fluttering, gliding, and perhaps foraging. But suddenly a swallow zipped down low over the clearing, violently chasing another one. One was making the typical alarm calls — high-pitched *tzee*, *tzee, tzee*s. The chase lasted only seconds as the couple skimmed low over the ground and then into the woods. In seconds the male was again perched on the apple tree, singing, and the female came flying in; he then flew down to the nest entrance and entered, and she perched silent on the apple tree.

Two hours later, near noon, the pair had been flying high over the clearing, independent of each other, when suddenly excited alarm calls sounded, and the male dived down and perched briefly in the box entrance. But then he flew up again to join three swallows, all near one another. In two minutes, all were gone. For the rest of the day the resident pair vacillated between perching on the apple tree, flying back and forth to their box, circling, leaving, and coming back.

The two made no progress on the nest over the next two days, until the morning of May 19, when, as was usual by then, they both arrived by around 6:30 a.m. and perched on the apple tree; then he flew into the box and peeked out, and she remained silent on the tree. But at 6:45 a.m. she flew up and flew *at* him as he perched in the nest-box entrance; she made a slight buzzing sound, as though to nudge him. He then left the box to fly up and perch at the top of the apple tree, while she swooped over the garden and started collecting grass. As he sang, she made trip after trip. But he left after she fell into a rhythm of grass collecting, and he stayed away for about ten minutes. She then stopped her work and parked in the nest-box entrance. As soon as he returned and sang from the tree again, she flew out and resumed her grass-carrying trips. He remained perched on the tree as she continued her work.

She had gathered only last year's dried grass and only in the garden, whereas the previous year's female had gathered only pine needles. This female used no pine needles. She also avoided fresh green grass, perhaps because she could not tear it up or access it as easily. Or didn't she like green grass? To find out, I salted the garden from one end to the other with freshly picked green grass blades, but she continued to take only the browned and the dried. However, after the green pieces had dried in the sun, she switched to taking them.

The dawn of May 20 was foggy after a rainy night, but then the sky cleared and the sun shone through passing clouds. No swallows came as I waited three hours for them, until finally the pair arrived, circled over the clearing for two minutes, made a few soft *chilk* calls, and then left.

Morning had always been the swallows' time for nest building and mating, but two hours later there was still neither sound nor sight of them. The birds' absence was surprising. It perhaps meant they were not concerned about competitors taking over their nest-box, or they had made so much progress on it that they were behaviorally ahead of the physiological schedule of egg production. The next day they showed up promptly at daybreak. For a half-hour they circled so high, they were only occasionally visible. Their flight was casual; it involved a lot of soaring, with intermittent flutters. But at 6:35 a.m., the male finally came down and zipped into the box. The female continued to soar as he flew to perch in the apple tree, where he sang until she finally came down (at 7:12 a.m.) and landed beside him. She then started gathering the day's first pieces of grass. She gradually picked up speed from one trip to another, until she completed one in as little as fifteen seconds. As before, all gathering trips were strictly to and from the garden. She took two more breaks, to sit beside him on the apple tree. When she joined him, he always stopped his singing, which was meant to induce interest in a particular nest-box or in nest building. He had succeeded on both counts, and singing was no longer necessary.

So far I had not seen a bird of any sort contest the pair's chosen nesting place, nor had there been any mating — in distinct contrast to all the other nestings I had observed. Chickadees were common visitors in the clearing, but the swallows ignored them; finches were also plentiful, as were mourning doves, which sometimes perched next to the swallows on the apple tree. Then, this morning at 6:30 a.m., a pair of eastern bluebirds flew into the clearing, and within a minute or two they were examining the nest-boxes. They entered most of them, from one end of the clearing to the other. Neither swallow budged to interrupt them.

By May 22 the nest looked like a haystack with a dent in the top; it looked ready for eggs.

But the female still took trip after trip between garden and nest to carry in more of the dried green grass. She spent as little as a second on the ground and two in flight, and usually five to fifteen seconds within the box. The male perched on the apple tree and occasionally sang. At about 9 a.m. both left. No egg had been laid, nor any new feathers brought in, but I expected egg laying soon. At 12:25 p.m. the *chee, chee, chee* alarm calls sounded, and three swallows neared the nest-box, one of them rapidly chasing after another. The two got tangled in midflight and fluttered toward the ground, but recovered themselves in the air, then quickly disappeared. None had returned by the end of the day. A third swallow made repeated appearances the next day, landing once at the box entrance. It hung there briefly, peeked in, then circled a couple of times before leaving.

On May 24 I woke up as usual to the reveille of the sapsucker drumming on the trunk of the apple tree. The clear sky on this wind-still dawn lacked swallows until 5:45 a.m., when one came circling very high. Fifteen minutes later a second joined it, and in another minute the male swooped out of the sky to hang at the nest-box entrance for a second and then fly up to his perch on the apple tree. But he had barely landed before flying on and circling; then the female flew to the box also. He merely looked in and then flew on. Ten minutes later they came back and repeated this pattern of movement. But at 6:18 a.m. he entered the box and then, from inside, repeatedly turned his head to look up to the empty sky, before leaving. Both returned forty-nine minutes later, at 7:12 a.m., and this time both entered the box. He left it after three minutes and she after five. I checked the box: no egg had been laid. And then I waited, and waited —

On this day the female swallow showed no interest in the garden, nor did she demonstrate any nest-building behaviors. It was time for mating — it should have been, and likely was, in process somewhere. During all my swallow watches of previous years, mating had been prominent and frequent, but now it did not occur near the nest nor within sight of the clearing.

The next morning (May 25) bumblebees had been up and humming, foraging from the wild apple-tree flowers; it was still almost dark at 4:40 a.m. But the pair of swallows did not arrive until two hours later, at 6:45 a.m., and they circled in silence, high over the clearing. Meanwhile the yellow-throated warbler had for an hour ceaselessly belted out one loud and vibrant *whitchety, whitchety* after another from the low shrubbery. A red-eyed vireo sang nonstop from the maple forest at the bottom of the clearing, along with, at intervals, a scarlet tanager. The song of the pipes-of-pan hermit thrush sounded from the depth of the dark, dense forest of pine, fir, and spruce, along with the songs of a winter wren and a blue-headed vireo. The tree swallows soared overhead for hours, in virtual silence. This quiet was puzzling; it seemed they were up to something. But more likely I was just projecting as I waited for the first egg, a benchmark in the nesting cycle. At 6:30 a.m., the nest contained a small white feather seemingly dropped on top of the nest bowl of deep dried grass. The pair came only twice more before noon, and each time the female brought a small piece of grass into the nest. At the end of the day, the nest still lacked an egg.

At 5:50 a.m. the next day there was no swallow in the sky over the clearing. But then, ten minutes later, one violently chased another, and the contestants disappeared over the forest to the west. I rushed to check the nest: there was still no egg. I then took a break in the cabin to escape the hordes of black flies. I did miss something: at 7:10 a.m., when I tapped the box to see if a swallow had returned, nothing flew out. Just to be sure, I opened the nest-box, and there sat the female on the nest! I quickly closed the box and left. She stayed inside, and I watched the box from inside the cabin through the window at my desk, expecting to see her leave and hoping then to find the first egg.

The female left twenty-seven minutes later, flying straight out and away, without circling once. I rushed out and rechecked the box: again, no egg had been laid. After that, no swallows came near for the rest of the day. As always, the yellow-throated warbler sang his two-second *whitchety, whitchety* every fifteen seconds, like clockwork. Ovenbirds continued to call at odd moments, and a male bluebird came again into the clearing, sang, and perched on the swallows' nest-box.

The next morning (May 27) the pair did not arrive until 6:20 a.m. They circled high and were silent, and within a minute one slipped into the nest-box, with the other entering right behind. Fifteen minutes later the male came up to the entrance and peeked out, and after a minute he left the box and started circling. Would he stay or leave? It seemed important to find out, because it would indicate his priorities. Keeping him in sight at all times became a marathon effort, one that I initially thought might last for 5 minutes, but ended up lasting 102 instead.

The male had remained virtually silent the whole time he circled alone, except when "the" merlin once again flew high over the clearing; then he immediately gave chase and made alarm calls. He approached this bird-predator but remained circumspect, coming no closer than about fifteen meters. The merlin did not swerve from its path; it was gone within seconds. But the swallow's emotion did not subside quickly; he kept on making alarm calls for the next fourteen minutes, at a gradually decreasing frequency.

I had expected the female to lay her egg and then rejoin the male in the air, and for me the wait seemed interminable. It also seemed odd that his singing, so prominent over the previous days, had stopped. Perhaps this was because the nest was finished and there was nothing for him to encourage her to do, except possibly lay the eggs. She now spent most of the morning in the nest, and when she finally peeked out of the box fifty-four minutes after she had entered it, she seemed calm and relaxed; she went back down into it after just a look around. But forty-five minutes later, at her second time up, she jerkily moved her head in all directions, casting an eye to the sky. Three minutes later, she flew out and joined the male, and then both left. My anticipation was mounting, somehow intensified by the black flies' torture. I ran to the nest-box, opened it, and saw the contents: no egg.

After we gain a little knowledge on a topic, we may pride ourselves on being able to make predictions, which is, after all, the main reason for going to the trouble of collecting facts. Often, though, the facts fail to make sense. Why, for example, didn't the swallows sleep at night in their now-cozy nest? Why had there been no mating? Why, if she was not going to lay an egg yesterday, or today, did she come into the nest-box in the morning and stay inside for over an hour and a half? And why did he circle nonstop, in silence, for that duration? Assuming a minimum flight speed of twenty miles per hour, he had just flown at least thirty miles. For what? What did he, or she, get out of this? If he was guarding the nest, why did he not stay close by, on his usual perch at the apple tree? And why there, and not on the black locust tree or the solar panel, which a previous pair had used exclusively?

By 5:15 A.M. on May 28 the yellow-throated warbler next to me in a bush had already been singing for a half-hour, and the hermit thrush still sounded in the woods. But the gray sky on this wind-still dawn showed no circling swallow, and it was silent. But both swallows arrived a half-hour later, and as before, circled high and in silence. Within five minutes of their arrival, the female again flew low over the clearing, made a turn, and dived directly into the nest-box. Within seconds, the male entered behind her.

Both stayed in the box. Why had he flown in too — to accompany her as she now presumably very soon would lay her first egg? His behavior could not be nest-box guarding; he could have accomplished that by perching on his apple tree, which offered a far better vantage point than the interior of the box. She surely did not need his encouragement to lay an egg. She had to do that by herself. We might propose that he was guarding his mate, but that too does not require this behavior. He had not helped in nest building and could not help with egg laying, and now his most prominent activity was flying in circles and staying with her in the nest. After twenty-five minutes he peeked out from the nest-box entrance for another minute, then shot out and resumed his high flying over the clearing. As he had done yesterday, he flew fast, as though trying to achieve a high speed. He would fly non-stop until 8:11 a.m., within four minutes of two hours. After that, as had happened yesterday, the two left for the rest of the day. In the meantime, I felt compelled to watch and try to catch every detail. Today's anticipated laying of the first egg was like an expected performance by an admired actor or musician. One does not walk out in the middle of the act, and so I kept watching.

The female had finally peeked out from the box entrance at 7:15 a.m., an hour after the male had resumed his sky flying. She had then occasionally glanced up, perhaps toward him, but for the most part she barely moved her head, and I presumed she would stay awhile. And she did: after fifteen minutes she dropped back down into the nest. But then she came up again at 7:38 a.m., and then went back down in another fifteen minutes, and then back up after a few minutes more. But this time she looked twitchy — she was jerking her head in quick motions. Then, at 8:11 a.m., she suddenly zipped out, and in an erratic, zigzagging swift flight, she headed low over the trees to the horizon. The male had meanwhile been up in the sky overhead the whole time, but now he joined her as though he had waited for her, and both left together. I rushed to the nest-box, lifted the front panel, and peeked in: *as expected*, one pearly-white egg lay in the deep cup of dried grasses. There was no feather near it. Of the nestings of tree swallows that I had seen thus far, this one had begun at the latest date. It also involved the least conflict over a nest-box. I expected the swallows' late start would have positive consequences; their baby rearing would occur during the warm, insect-rich summer season.

18

LATECOMER'S
ADVANTAGE — 2018

As usual, one egg was laid per day in the morning. The fifth and last one was in the swallows' nest on June 1. Two days later the female was peeking out of the nest-box entrance in the evening, and she flushed briefly from the nest as I looked in. I saw that the nest still held five eggs, but now also six feathers. No parasite egg had been inserted by this time, when one might have been effective. Presumably and inexplicably, mating had been occurring elsewhere, the eggs would be fertile, and babies would fledge. I now continued as before to watch the nest daily from a distance and to observe responses to feathers.

Yes, it was feather time now. On June 6, the male dropped three small light-colored feathers onto the ground in front of the nest-box. They were released from his bill, as if by accident, while he was entering the nest. The female did not take them. Was she not ready to do so, had they been delivered improperly, or did she find them unsuitable? To find out, the next morning I placed three long (twelve-centimeter) white feathers on the ground with the others, while she was in the box. Ten minutes later, as she was peeking out of the entrance and preparing to leave, she kept turning her head and repeatedly looking down to them. She then flew out of the box, took one of the long ones I had provided, circled the clearing several times with it, and took it into the box. She then also picked up the other two large ones, flying with each of these directly into the box, with no circling. The three small ones her mate had brought were left behind. His had apparently been rejected.

As at other nests, the gathering of feathers started in earnest with incubation and continued throughout its duration. The preferred long feathers helped hide the eggs, but as I've said earlier, if they were brought simply to camouflage the eggs to deter cuckolds, adding the feathers during incubation would make little sense, except perhaps to hide them when the female necessarily has to leave the nest during incubation to find food for herself (tree swallow males do not feed their mates during this time, as do most other birds).

There was little new to be expected from my continuous monitoring of the feather gathering and the behaviors accompanying it. Whether a foreign egg might be added after the nest was finished and while feathers were being inserted was still at least a possibility, and so I kept checking. What I found, however, was more surprising and more interesting than an extra egg or more feathers. I had checked the nest contents every day, and I noted that all five eggs hatched on June 16. Over the next several days all seemed healthy, but on June 21 there was one less baby. (And then on June 24 still another was missing.) Once again, now for the third nesting cycle, a baby was mysteriously gone. Despite my careful watching, I had no clue as to how this was done or who did it.

As I carefully surveyed the new babies on the 16th, I saw that they were snugly packed into the nest, and all seemed healthy. At about 4 p.m., right after the pair left, a swallow appeared and made four approach flights by the box, but it did not dive in, as the parents had both done routinely before. The extreme hesitancy of this female swallow at the nest-box entrance indicated that she was an extra-pair individual. On the fifth approach she merely hung at the nest-box entrance and peeked in, flew off, then returned, and she repeated this behavior three times before finally slipping into the nest-box. No adult swallow was inside, because normally a flyby served as a signal for the mate to fly out. Three minutes later, the male returned, but still the bird did not fly out. The male hung briefly at the box entrance, and then this new bird slipped out and flew off. It was either a female or a bird in juvenile garb. He ignored it.

No new feathers were added to the nest over the next few days. Then, on June 21, at 6 a.m., I again placed (six) long, especially fluffy white feathers on the ground in front of the nest-box. Feather gathering had stopped, but might the swallows be tempted to pick up "super" feathers so readily available? They were; at 9 a.m. all six feathers were inside the nest-box. I then put down seven more (two black, three gray, two white). By 11 a.m., both the female and the male had carried all of these into the nest. The last feather picked up, at 10:55 a.m., was the smallest of this lot, and as the female took this one, she did not fly directly back into the box but carried it high into the air, where she dropped it. As it started drifting down, she turned, caught it, and dropped it again, repeating this sequence eight times before starting to enter the box. At this point the feather apparently slipped from her bill and drifted down. She instantly turned back, caught it, and this time dropped it twice more in the air before returning to the box with it. This behavior, similar to actions that occurred before feather gathering had begun, suggested that the bird was reverting to play; function and immediate utility were not at issue now.

Over the next two days (June 22–23), I placed ten long feathers on the ground by the nest-box. They were all taken in within twenty minutes. I then placed five more on the ground; they were picked up within an hour. As I photographed the now well-feathered nest, with its four half-grown young, I was hugely surprised to see one baby that still looked pink and was much smaller than the rest. The next day this smaller one was missing; the original clutch of five eggs had now been reduced to three babies. There was no sign of a predator in either instance of brood reduction.

The next morning at 10:40 a.m., under an overcast sky, there were no swallows around. But at nearly 11 a.m. the female dived into the box and came out almost immediately, flying west. At 11:12 a.m. she was again in and out, but instead of flying off low and fast in one direction, she circled high and then dropped a small feather; she had apparently taken it out with her when she left the nest. As it drifted down, she swooped, caught it, and repeated this sequence four times; then she flew off out of sight. Her taking a feather from her nest and then playing with it made me wonder. Were the removals of fecal waste sacs, eggshells, and, yes, babies somehow behaviorally linked? Was the female swallow now in the mode of removing from the nest anything that did not belong there?

Tossing two babies out of the nest seems cruel to us. But it is an anthropomorphism to compare what a swallow does to a human performing the equivalent action. We are equipped with different emotions that help preserve our lives. Unless extreme circumstances dictate otherwise, the automatic and therefore preferred response to a challenge is based on emotion. In the case of the swallows, which almost routinely experience starvations, this baby removal may have been merciful. Behavior is directly guided by emotions or feelings. However, the adaptive behavior that produces the best long-term results in a given situation evolves not by means of feelings, but by the creation of emotions that induce the behavior.

At 11:44 a.m., the pair of swallows flew high over the clearing, but now a third was peacefully flying with it. Then one of the three dropped low and flew past the nest-box, repeating six passes in a row. Why did it not dive in? From its flight alone I could not determine whether it was male or female, but suddenly a female did land at the entrance — she was holding an insect in her bill, and she hung there and looked in. And then in an instant, a shiny blue male landed beside her and seemed to try to shove her aside; he then grabbed the fly out of her bill and slipped into the box with it. This scenario made no sense. But in seconds it was over, and the female slipped into the box after the male left. She stayed in the nest for two minutes before coming back up to the entrance, where she perched another minute or so before leaving. There was much more going on here with diverse facts that could not be dismissed as meaningless random events. Statistical rules can be derived on the population level, but it is the behavior of individuals that make those rules, not the other way round.

The next day (June 25) I removed all the feathers from the nest and put ten on the ground in front of the nest-box. Three days later, not one had been taken, but the three young in the nest were now rapidly growing and starting to feather out.

THE WEATHER CONTINUED to be ideal, and the three baby swallows were now thriving; they were rapidly becoming feathered. Unlike the young in some previous years, none of these three nestlings were appearing at the nest entrance. But on this afternoon (July 5), one of the hottest of the summer, with temperatures rising to 90 degrees F, I heard *cheep, cheep* calls, and the female dived at my head when I approached the nest-box. I stood still, and she again and again flew high and then dived directly at my head, missing it by mere centimeters. I backed off as I continued past the box for a quick dip in the nearby brook. Her instinct to defend her babies had clearly been awakened, full force. The defensive behavior signaled that fledging was about to happen, and I expected it to be a drawn-out process. So I took that cool dip in the brook. A mistake: when I came back within the half-hour, all was silent. The box was empty — the young had fledged. I never saw one of them again; they had left not just the nest-box but also the clearing and the area. Neither parent ever came back, not even for a single flyby. The three young had been adequately fed; they had been strong flyers upon leaving the nest.

Similarly, in 2019 — a season outside the scope of this book — there was also a reduction in nestlings, from four eggs to only two babies at fledging. But there was a very noticeable change in pattern of earlier fledglings on one front, but a congruence with this one: there was no active "luring" behavior by the male or the female at fledging. This time the female seemed to casually enter the nest box as the babies were ready to fledge, but while she was still down in it, the last baby simply hopped up and into the nest entrance, where the female joined it, then quietly yielded to let the baby slip by her and jump out into full flight. Both were soon out of sight in the distance.

Now, with the 2018 babies fledged, I pulled out all the nest-box contents. This time the nest was clean — there was hardly any fecal matter in it. Since none of the three had blocked the entrance to try to intercept food, the parents had entered the box at each feeding, and they had in the process been able to access and remove the feces. I checked the nest-box contents carefully. Unlike many nests of the past, this one contained no remains of a dead swallow. This had been a successful nest, placing little stress on the young. They were well prepared for life, starting out with an inheritance of strong flight muscles and undoubtedly a layer of fat, which would allow them to fly fast and far. They had deviated from the statistical norm, which may on average be the best, but evolution works on individuals. The norm is created by individual choices.

From this day on, as in other years, the tree swallows disappeared as if swept from the country. In my mind I would follow their journey to winter roosts in the swamps and wetlands along the Atlantic and Gulf coasts, to gatherings of millions that funnel out of the evening sky, a mass of swallows like smoke descending to enter the reed thickets where they rest overnight. Tree swallows are birds of the sky. I will miss them, and I wish them a good journey, with plenty of bayberries to eat on the Maine coast, lots of bugs in the South, and as always, safe travels back home.

POSTSCRIPTS

A CONCLUDING EXPERIMENT

Much of the literature on tree swallows is based on studies involving a sample size of dozens to hundreds of swallows and nest-boxes arranged in a grid pattern, in an open area of unspecified size. In contrast, only one pair of tree swallows nested in my half-hectare clearing in the woods per year, despite the fact that there were always nine nest-boxes available. These differences in scale remind me of a pair of ravens here, who defended their turf well until they were overrun by a hundred or more vagrants attracted by a moose carcass nearby. At a certain point, a resident pair's defense becomes useless; after they chase one intruder off, a dozen more come in. I also remember how one red squirrel generally chased off all others from our bird feeder until the winter of 2018–19. Then, following a recent great year for mast (tree nuts fallen to the ground), the squirrel population shot up spectacularly. Dozens of red squirrels overran the premises, and they no longer chased after one another. All defense of territory ceased. Think of humans living on separate home-steads and preventing strangers from setting up housekeeping nearby; in a city it would be impossible to defend the same amount of space. Similarly, swallows adapted to living in open space as opposed to forest would likely feel crowded if enclosed in a small clearing.

To test the effects of crowding, I set up an experiment. I put up two identical nest-boxes in a clearing similar in size to ours, but six kilometers away: in the three years I did this, only one pair of sparrows used a box. In another experiment, in a nearby (within five kilometers) 4.5-hectare field, I spread eighteen identical new nest-boxes for tree swallows, placing them evenly twenty-five meters apart, along a straight field edge bordering on forest. In the first spring (2016) the tree swallows that arrived there took an immediate interest in the new nest-boxes. Nests (made of grass carried into the boxes) were started in half of the eighteen available nest-boxes. However, only four boxes fledged young. The second year's result was nearly identical, with ten nests started, but again only four of the boxes fledged young.

It seemed odd that at least twice as many nests were started than were finished and produced young. Such behavior hints at a disturbance related to crowding — confusion as to which box is which. In one case this was directly obvious: in two of the identical boxes separated by twenty-five meters along the field edge, one nest held only a thick layer of grass for the nest proper, and the nearby box had no grass but a nest lining made up of a single feather instead. Furthermore, both nests had been deserted after about half a clutch of eggs (two and three eggs, respectively) had been laid in each; the same pair of swallows had used the two boxes simultaneously, without completing their nesting in either. The other three unsuccessful starts (also deserted) had nests made of grass only. If the swallows were indeed confused about which nest was their *own* nest among the identical boxes, it's likely that they also inadvertently entered other swallows' nest-boxes. This must have triggered conflict. Such entrances can be mistaken for aggression or attempted nest-box takeovers.

The nest-boxes were all made from freshly sawed, unweathered pine board, with identical round entrances, one and a quarter inches in diameter and facing in the same direction. The boxes were spaced evenly apart. What would happen if the boxes were made to look different or were placed differently?

For a follow-up test, I kept the same number of boxes in the same positions but changed their appearance. I painted a third of them white and another third black; the rest remained natural pine, with no added color. The swallows used boxes of all three kinds, and this time there were only two, rather than five, unfinished nest starts, and twice as many (eight) nest-boxes fledged young. This result supports the idea that nest-box confusion had been a factor in the swallows' behavior previously.

During this time, in the second and third springs, near the same number of boxes (twenty-one) were simultaneously set up at two other nearby sites (within three and four kilometers of the previous set). These boxes were identical to the first lot, made in the same design and by the same person. Although I anticipated that the local tree swallows might not need so much new housing, and that possibly no swallows would use these boxes, it turned out precisely the opposite.

I had spaced these boxes at least twice as far apart. Additionally, they were located near environmental features such as a tree, a fence, bushes, and the edge of a field. This time there were no unfinished nest starts, and over these three years, eleven, seventeen, and seventeen, respectively, of these twenty-one boxes fledged young.

There had been no shortage of swallows with respect to the available nest-boxes in the wide-open areas. With several times more sky space, the extraordinary result was that almost every available nest-box was used (other regular residents of the boxes included bluebirds, black-capped chickadees, and a deer mouse). These data show that there had been a large housing shortage all along, despite the many unused boxes in my clearing. Unlike most other swallow species, tree swallows are not communal nesters. In a housing shortage they may be forced into a semi-communal nesting space despite their highly territorial nature; any behavior they display under these conditions is likely to be biologically unnatural. Nevertheless, in the wide-open areas they acted at times communal, in that if one swallow was alarmed while I was examining a box with a clutch of young, all the neighbors arrived and circled around me.

These results help explain what had seemed to be enigmatic behavior: the annual nesting of just one pair of swallows in the defined space of the forest clearing. This was consistent with the hypothesis that tree swallows are not, like other swallows, tolerant of neighbors. When they can, the male and the female tree swallows control the sky around their nest cavity. Their nesting place is their most valuable resource. But when many nest-boxes are placed together, conflict ensues, especially if the boxes are indistinguishable. Dissolved boundaries lead to unintended nest invasions. The efforts of individuals mean little amid the confusion of the crowd. Thus, the basis of much research into tree swallows — those artificial colonies of identical nest-boxes evenly spaced in grids in a field — can reveal only a small part of the world of the tree swallow.

Having some neighbors is normal in the tree swallow's natural habitat of beaver bogs. In the nearest one, about twice the size of my clearing and located a mile and a half to the south, many dead trees stand in the water. The water protects nests from squirrels and other predators. Great blue herons nest in those trees. Woodpeckers of potentially five species and two kinds of nuthatches excavate their nest holes in the partially decayed trees. In the winter of 2017–18 (when I could walk around the flooded bog on the ice of the dammed water), I located at least a dozen potentially useful nesting cavities. However, there was no way for me to know if they had been used, and I realized then that, in the nesting season, neither might a swallow be able to tell if a particular cavity had already been claimed. Each nest site in a beaver bog is spaced apart and marked by immediately local and highly variable features.

However, a tree swallow searching for a nest site is unlikely to know beforehand if a particular hole is occupied until it has entered it. Being able to see showy, long white feathers at the cavity could signal this fact instantly — the site is occupied; it has defenders. This nest-hole searcher could spare itself the energy it takes to fight, and instead choose to hunt for and examine another nest site. Finding a cavity without the white-feather signal of occupancy, a swallow may become enthused and claim it, at this point perhaps willing to invest in a prolonged battle, if needed. Time is of the essence; avoiding a battle can open up a win-win option both for newcomers and owners of nest holes.

Cliff and barn and bank swallows use no feathers, or sometimes only several small white feathers, to line their nests. Their nest sites are far less valuable; they can be built at almost any spot anywhere, and hence are not worth fighting for, nor labeling with a sign of ownership.

Eight years of observing "my" swallows' behaviors related to white feathers yielded both fascination and frustration. It was these concluding experiments, with the added nest-boxes, that in the end yielded a richly satisfying "answer" concerning tree swallows' affinity for white feathers. However, as much as this idea appeals, it needs to be tested. And even if proved, it would not preclude the possibility of other selective pressures (such as the need for insulation or hiding the eggs from nest parasites or predates) having simultaneously contributed to the use of feathers.

TREE SWALLOWS IN WINTER

From the moment the last fledged baby tree swallow leaves our clearing, no more swallows will be sighted there until the next spring. Yet the swallows' epic life is just beginning, and it spans the continent. For several days after they fledge in early July, small groups of tree swallows forage over local swamps and meadows. A small swarm may be sighted here or there for a short time, but then they are gone.

One fall I received an email from my friend Professor Nathaniel Wheelwright of Bowdoin College, showing a photograph, by his student Liam Taylor, of a mass of tightly packed tree swallows perched near the Atlantic shoreline in a marsh at Biddeford Pool in southern Maine. The birds were feeding on their winter food plant, the northern bayberry, or wax myrtle (*Myrica pensylvanica*). Bayberries are small, tightly clumped, pale blue, and wax-covered. Early American colonists boiled them down to skim off the wax for making candles. This wax on the berries helps preserve them through the winter, and their blue berries provide sustenance for the swallows. (A second similar bayberry species, which hybridizes with the first, is the southern bayberry, *M. cerifera*, which extends south along the Gulf Coast and to Central America.)

In Taylor's photograph, the birds are so densely packed, they practically touch one another, aggregated into the hazy horizon in the picture. Among the swallows in the foreground, there appears to be a near-equal mixture of shiny blue (adults) and brown (mostly juvenile) birds. Taylor and Wheelwright saw several such swarms and estimated "at least 2000" birds were present on the early afternoon of September 6, when the photograph was taken. However, by my count, that is an extremely conservative estimate, if one considers the barely visible masses of birds in the distance as well.

As the season progresses, these aggregations of tree swallows (joined, occasionally, by other swallows, such as bank swallows) move southward to new berry-feeding grounds, where still more join them — a continuing spectacle. On September 30, 1991, Robert Finch (as he recorded in his essay in *The Nature of Nature*) saw them a hundred miles farther south, at Cape Cod Bay. He wrote, "In the dunes between the beach and the harbor was an amazing sight. Over a thousand iridescent blue-green tree swallows were flocking among the bayberry and goldenrod, buzzing, perching, then taking off in sudden mass bursts of flight, only to turn around and group again in a kind of constant dynamic cohesiveness, their white bellies flashing, morning light glinting off their backs."

Recently (2016) the ornithologist Paul R. Spitzer, during his work on the restoration of ospreys, observed the tree swallows where the Choptank River flows into Chesapeake Bay. Huge numbers start aggregating in roosts in September and remain through October in the phragmites and cattails, as they do also on Great Island Marsh on the Connecticut River estuary. He wrote me this: "The huge annual roost is forming every evening down at the Big Bend of the Choptank River, in large aprons of marshes on the Trappe side. Around sunset they stream in from all directions, circle high and wide for roughly half an hour or more until dark approaches, then swirl together and downwards like a dark tornado funnel that feeds the reeds of the waiting marsh below. [There are also] wood duck flights, the wind, the light. It is fabulous, very celebrating, very spiritual."

Later on, in the winter, the tree swallows swarm still farther south, and at dusk they blacken the skies. Roger Tory Peterson, the maven of bird watchers, had marveled at the experience and described it like this: *"For sheer drama, the tornadoes of Tree swallows eclipsed any other avian spectacle I have ever seen."*

Finally, in early January, the swallows arrive at the southwestern coast of Florida, near Sarasota, Rotunda West, and Ruskin. A spectacular recent YouTube video (September 1, 2017) shows those throngs of millions. But the most amazing YouTube videos were two created by Mark H. Vance, on January 5, 2011, and January 4, 2012, in Sarasota. The videos show a mind-boggling three million swallows performing their flight murmurations in unison. On and on they maneuver, in seemingly endless braiding, in fantastic, chimerical shape-changing formations that undulate in the sky in all directions. The mass of swallows changes not just in shape but also from dark to light, as the millions angle in unison in their flight. They were filmed as they were coming in to roost for the night.

ANOTHER SERIES of YouTube videos of the tree swallows (also at the Florida Gulf Coast, but at Ruskin, near Tampa), recorded on March 14, 2012, and February 23 and March 6, 2017, shows them in flight over the protected marshes near Cockroach Bay. The birds were again present in the millions, spiraling in funnels out of the sky to land in the marsh grass, where they stay for the night. The swallows look like clouds coalescing in tornado funnels, spiraling down one after another.

From the wetlands of Florida, some of the tree swallows then move on along the Gulf Coast to coastal Louisiana, where spectacular roosts have been reported in sugarcane fields. From late October through December, similar if not larger throngs overwinter in the wetlands of coastal Louisiana. It is likely that these birds had come in steps, with some coming south from extended stops along the Mississippi River drainage, and some continuing through Central America. Near Vacherie, Louisiana, they use sugarcane fields as apparent substitutes for reed beds for overnighting, an estimated five million to a roost.

Although the tree swallows are the slowest of any songbirds in their migration south, since they stop to feed on bayberries along the way, they are one of the earliest to return north, a journey taken again in steps — it takes the population only one to two months to complete the journey. Presumably there are now fewer staging areas stocked with food, so they need to hurry. On the other hand, they can expect even less food when they get back to their old nest sites, and the weather may be unpredictable where they were born. But they must hurry in any case, because there are very few nest sites, and very few of those will be vacant.

THE LITERATURE

The scientific literature on swallows, especially tree swallows, is extensive. Already in 2003, when Jason Jones published an overview of the research invested in this species, he found that four hundred manuscripts had been produced over the past twenty-five years. By 2017 this amounted to about five hundred studies. The topics range widely, and the extent and breadth of studies led Jones to describe the tree swallow as a "model" study organism. However, whereas cliff swallows are arguably a model study organism for the study of social behavior, tree swallows have become primarily a model study organism for mating behavior, specifically extra-pair copulation. This topic is represented in about sixty of these publications on tree swallows. For this list, I have here pruned the literature to minimize the offerings on mating and to cite research relevant to my questions and observations.

Ardia, D. R. 2013. The effects of nest-box thermal environment on fledging success and haematocrit in tree swallows. *Avian Biology Research* 6: 1–6.

Ardia, D. R., J. H. Perez, E. K. Chad, M. A. Voss, and E. D. Clotfelter. 2009. Temperature and life history: Experimental heating leads female tree swallows to modulate egg temperature and incubation behavior. *Journal of Animal Ecology* 78: 4–13.

Ardia, D. R., J. H. Perez, and E. D. Clotfelter. 2006. Nest box orientation affects internal temperature and nest site selection by tree swallows. *Journal of Field Ornithology* 77: 339–44.

Barber, C. A., and R. J. Robertson. 1998. Homing ability of breeding male tree swallows. *Journal of Field Ornithology* 69: 444–49.

Barber, C. A., and R. J. Robertson. 1999. Floater males engage in extrapair copulations with resident female tree swallows. *Auk* 116: 264–69.

Bitton, P.-P., and R. D. Dawson. 2008. Age-related differences in plumage characteristics of male tree swallows *Tachycineta bicolor:* Hue and brightness signal different aspects of individual quality. *Journal of Avian Biology* 39: 446–52.

Brown, C. R. 1984. Laying eggs in a neighbor's nest: Benefits and costs of colonial nesting. *Science* 224: 518–19.

Brown, C. R., M. B. Brown, P. Pyle, and M. A. Patten. 2017. Cliff swallows (*Petrochelidon pyrrhonota*), version 3.0. In *The Birds of North America* (P. G. Rodewald, ed.). Cornell Lab of Ornithology, Ithaca, NY. https://doi.org/10.2173/bna.clswa.o3.

Brown, C. R., and M. B. Brown. 1984. Behavioral dynamics of interspecific brood parasitism in colonial cliff swallows. *Animal Behavior* 37: 777–96.

Burtt, E. H. 1977. Some factors in the timing of parent-offspring recognition in swallows. *Animal Behavior* 25: 231–39.

Butler, R. W. 1988. Population dynamics and migration routes of tree swallows, *Tachycineta bicolor,* in North America. *Journal of Field Ornithology* 59: 395–402.

Chaplin, S. B., M. L. Cervenka, and A. C. Mickelson. 2002. Thermal environment of the nest during development of tree swallow (*Tachycineta bicolor*) chicks. *Auk* 119: 845–51.

Chek, A. A., and R. J. Robertson. 1994. Weak mate guarding in tree swallows. *Ethology* 98: 113.

Clotfelter, E. D., L. A. Whittingham, and P. O. Dunn. 2000. Laying order, hatching asynchrony, and nestling body mass in tree swallows *Tachycineta bicolor. Journal of Avian Biology* 31: 329–34.

Coady, C. D., and R. D. Dawson. 2013. Subadult plumage color of female tree swallows (*Tachycineta bicolor*) reduces conspecific aggression during the breeding season. *Wilson Journal of Ornithology* 125: 348–60.

Conrad, K. F., P. V. Johnston, C. Crossman, B. Kempenaers, R. J. Robertson, N. T. Wheelwright, and P. T. Boag. 2001. High levels of extra-pair paternity in an isolated, low-density, island population of tree swallows (*Tachycineta bicolor*). *Molecular Ecology* 10: 1301–8.

Crowe, S. A., O. Kleven, K. E. Delmore, T. Laskemoen, J. J. Nocera, J. T. Lifjeld, and R. J. Robertson. 2009. Paternity assurance through frequent copulations in a wild passerine with intense sperm competition. *Animal Behavior* 77: 183–87.

Dakin, R., A. Z. Lendvai, J. Q. Ouyang, I. T. Moore, and F. Bonier. 2016. Plumage colour is associated with partner parental care in mutually ornamented tree swallows. *Animal Behavior* 111: 111–18.

Dawson, R. D. 2008. Timing of breeding and environmental factors as determinants of reproductive performance of tree swallows. *Canadian Journal of Zoology* 86: 843–50.

Dawson, R. D., C. C. Lawrie, and E. L. O'Brien. 2005. The importance of microclimate variation in determining size, growth and survival of avian offspring: Experimental evidence from a cavity nesting passerine. *Oecologia* 144: 499–507.

Dawson, R. D., E. L. O'Brien, and T. J. Mlynowski. 2011. The price of insulation: Costs and benefits of feather delivery to nests for male tree swallows *Tachycineta bicolor*. *Journal of Avian Biology* 42: 93–102.

Dunn, P. O., and S. J. Hannon. 1992. Effects of food abundance and male parental care on reproductive success and monogamy in tree swallows. *Auk* 109: 488–99.

Dunn, P. O., J. T. Lifjeld, and L. A. Whittingham. 2009. Multiple paternity and offspring quality in tree swallows. *Behavioral Psychology and Social Biology* 63: 911–22.

Dunn, P. O., and R. J. Robertson. 1994. Do males exchange feathers for copulations in Tree Swallows? *Auk* 112: 1079–80.

Dunn, P. O., and L. A. Whittingham. 2005. Radio-tracking of female tree swallows prior to egg-laying. *Journal of Field Ornithology* 76: 259–63.

Heinrich, B. 1989. *Ravens in Winter*. Summit Books of Simon & Schuster, NY.

Heinrich, B. 2004. *The Geese of a Beaver Bog*. HarperCollins, NY. (See page 217.)

Heinrich, B. 2015. Tree swallows' feather-lining their nest: An antiparasitizing strategy? *Northeastern Naturalist* 22 (3): 521–29.

Heinrich, B. 2010. *Nesting Season: Cuckoos, Cuckolds, and the Evolution of Monogamy*. Harvard University Press, Cambridge, MA. (See page 337.)

Hussell, D.J.T. 1983. Age and plumage color in female tree swallows. *Journal of Field Ornithology* 54: 312–18.

Jones, J. 2003. Tree swallows (*Tachycineta bicolor*): a new model organism. *Auk* 120: 591–99.

Kilham, L. 1980. Assemblages of tree swallows as information centers. *Florida Field Naturalist* 8: 26–28.

Lack, D. 1947. The significance of clutch size. *Ibis* 89: 302–52.

Leonard, M. L., and A. G. Horn. 2001. Begging calls and parental feeding decisions in tree swallows (*Tachycineta bicolor*). *Behavioral Ecology and Sociobiology* 49: 170–75.

Leonard, M. L., A. G. Horn, and A. Dorland. 2009. Does begging call convergence increase feeding rates to nestling tree swallows *Tachycineta bicolor*? *Journal of Avian Biology* 40: 243–47.

Lombardo, M. P. 1986. Attendants at tree swallow nests. 1. Are attendants helpers at the nest? *Condor* 88: 297–303.

Lombardo, M. P. 1988. Evidence of intraspecific brood parasitism in the tree swallow. *Wilson Bulletin* 100: 126–28.

Lombardo, M. P. 1995. Within-pair copulations: Are female tree swallows feathering their own nests? *Auk* 112: 1077–79.

Lombardo, M. P., R. M. Bosman, C. A. Faro, S. G. Houtteman, and T. S. Kluisza. 1995. Effect of feathers as nest insulation on incubation behavior and reproductive performance of tree swallows (*Tachycineta bicolor*). *Auk* 112: 973–81.

McCarty, J. P., and D. W. Winkler. 1999. Foraging ecology and diet selectivity of Tree Swallows feeding nestlings. *Condor* 101: 246–54.

Michaud, T., and M. Leonard. 2000. The role of development, parental behavior, and nestmate competition in fledging of nestling tree swallows. *Auk* 117: 996–1002.

Moeller, A. P. 1987. Intraspecific nest parasitism and anti-parasite behavior in Swallows, *Hirundo rustica*. *Animal Behavior* 35: 247–54.

Moeller, A. P. 1991. The effect of feather nest lining on reproduction in the swallow *Hirundo rustica*. *Ornis Scandinavica* 22: 396–400.

Murphy, M., T. Armbrecth, E. Vlamis, and A. Pierce. 2000. Is reproduction by Tree Swallows cost free? *Auk* 117: 902–12.

Nebel, S., A. Mills, J. D. McCracken, and P. D. Taylor. 2010. Declines of aerial insectivores in North America follow a geographic gradient. *Avian Conservation and Ecology* 5: 1–14.

Quinney, T. E., D.J.T. Hussell, and C. D. Ankney. 1986. Sources of variation in growth of tree swallows. *Auk* 103: 389–400.

Rendell, W. B. 1993. Intraspecific killing observed in tree swallows, *Tachycineta bicolor*. *Canadian Field Naturalist* 107: 227–28.

Rendell, W. B., and R. J. Robertson. 1989. Nest-site characteristics, reproductive success, and cavity availability for tree swallows breeding in natural cavities. *Condor* 91: 875–85.

Rendell, W. B., and R. J. Robertson. 1990. Influence of forest edge on nest-site selection by Tree Swallows. *Wilson Bulletin* 102: 634–44.

Rendell, W. B., and R. J. Robertson. 1994. Defense of extra nest-sites by a cavity nesting bird, the tree swallow, *Tachycineta bicolor*. *Ardea* 82: 273–85.

Robertson, R. J., and W. B. Rendell. 1990. A comparison of the breeding ecology of a secondary cavity nesting bird, the tree swallow (*Tachycineta bicolor*), in nest-boxes and natural cavities. *Canadian Journal of Zoology* 68: 1046–52.

Robertson, R. J., and W. B. Rendell. 2001. A long-term study of reproductive performance in tree swallows: the influence of age and senescence on output. *Journal of Animal Ecology* 70: 1014–31.

Robertson, R. J., and B. J. Stutchbury. 1988. Experimental evidence for sexually selected infanticide in tree swallows. *Animal Behavior* 36: 749–53.

Robertson, R. J., B. J. Stutchbury, and R. R. Cohen. 1992. Tree Swallow. In *The Birds of North America*, No. 11 (A. Poole, P. Stettenheim, and F. Gill, eds.). Academy of Natural Sciences and American Ornithologists' Union, Washington, DC.

Shutler, David, et al. 2012. Spatiotemporal patterns in nest-box occupancy by tree swallows across North America. *Avian Conservation and Ecology* 7: 3. http://www.ace-eco.org/vol7/iss1/art3/

Stapleton, M. K., and R. J. Robertson. 2006. Female tree swallow home-range movements during their fertile period as revealed by radio-tracking. *Wilson Journal of Ornithology* 118: 502–7.

Steven, D. D. 1980. Clutch size, breeding success, and parental survival in the Tree Swallow (*Iridoprocne bicolor*). *Evolution* 34 (2): 278–91.

Stoddard, P. K., and M. D. Beecher. 1983. Parental recognition of offspring in the Cliff Swallow. *Auk* 100 (4): 795–99.

Stutchbury, B. J. 1998. Female mate choice of extra-pair males: Breeding synchrony is important. *Behavioral Ecology and Sociobiology* 43: 213–15.

Stutchbury, B. J., and R. J. Robertson. 1987. Signaling subordinate and female status: Two hypotheses for the adaptive significance of subadult plumage in female tree swallows. *Auk* 104: 717–23.

Thys, B., R. Pinxton, T. Raap, G. DeMeester, H. F. Rivera, and M. Eens. 2017. The female perspective of personality in a wild songbird: Repeatable aggression relates to exploratory behavior. *Scientific Reports* 7, article number 7656. doi:10.1038/s41598-017-08001-1.

Vleck, C. M., and D. Vleck. Hormones and regulation of parental behavior in birds. www.sciencedirect.com/science/article/pii/B9780123749291100071.

Weatherhead, P. J. 1988. Adaptive disposal of fecal sacs. *Condor* 90: 518–19.

Weaver, H. B., and C. B. Brown. 2004. Brood parasitism and egg transfer in Cave Swallows (*Petrochelidon fulva*) and Cliff Swallows (*P. pyrrhonota*) in South Texas. *Auk* 121: 1122–29.

Wheelwright, N. T., J. Leary, and C. Fitzgerald. 1991. The costs of reproduction in Tree Swallows (*Tachycineta bicolor*). *Canadian Journal of Zoology* 69: 2540–47.

Whittingham, L. A., and P. O. Dunn. 2001. Females' responses to intraspecific brood parasitism in the tree swallow. *Condor* 103: 166–70.

Whittingham, L. A., P. O. Dunn, and R. J. Robertson. 1995. Do males exchange feathers for copulations in tree swallows? *Auk* 112: 1079–80.

Whittingham, L. A., and H. Schwabl. 2002. Maternal testosterone in tree swallow eggs varies with female aggression. *Animal Behavior* 63: 63–67.

Whittingham, L. A., P. D. Taylor, and R. J. Robertson. 1992. Confidence of paternity and male parental care. *American Naturalist* 139: 1115–25.

Windsor, R. L., J. L. Fegely, and D. R. Ardia. 2013. The effects of nest size and insulation on thermal properties of tree swallow nests. *Journal of Avian Biology* 44: 305–10.

Winkler, D. W. 1993. Use and importance of feathers as nest lining in tree swallows (*Tachycineta bicolor*). *Auk* 110: 29–36.

INDEX

American goldfinches, 34, 51

American robins
hatching of eggs of, 154
observation of, 5, 34
song of, 51, 106, 107, 131, 148

bank swallows, 10
barn swallows, xii–xiii, 9–10, 90, 187, 213–14
barred owls, 51, 54, 116
bayberries, 215
beavers, 114
beetles, burying/sexton, 86–88
blackbirds, red-winged, 49, 51
black-capped chickadees, 77, 110
blue jays, 2, 6, 34, 176, 191
bluebirds, 110, 197, 212
eastern, 51, 194–95

blue-headed vireos, 51, 54, 90, 196
broad-winged hawks, 4
brood reduction, 89–91, 204–5. *See also* nest parasitization
brown drake (*Ephemera simulans*), 80
burying beetles, 86–88
butterflies, mourning cloak, 106

cedar waxwings, 5
chestnut-sided warblers, 18, 34
chickadees
black-capped, 77, 110–11
incubation and, 154
investigating nest-box, 82, 109, 143, 169–70
nest parasitization and, 46, 131, 135, 136, 139–40

chickadees *(cont.)*
  nest-box competition and,
    110–11, 112
  song of, 106, 122
  using nest-boxes, 1, 212
  visits from, 191, 194
cliff swallows, xii–xiii, 10, 64,
    187, 213–14
Cooper's hawks, 3, 4
cowbirds, 39, 106, 172
crowding, experiment on,
    210–14
crows, 176, 178
cuckoos, 39, 90

duck feathers, 11
ducks, 49

eagles, 90, 91
eastern bluebirds, 51, 194–95.
    See also bluebirds
eggs
  coloration of, 39
  hatching of, 11–12, 22, 102,
    139, 142–43, 167–68,
    169–70, 202
  incubation of, 1, 32, 44,
    71–73, 100, 135
  laying of, 11, 30, 38–40,
    42–46, 61–62, 98–100,
    130–32, 134, 162–65,
    166, 200, 201
  nest parasitization and,
    36–37, 39–40, 121–22,
    131, 133–35, 172
  number of, 89–91

emotions, 186, 205
English sparrows, 110
Ephemera simulans (brown
    drake), 80
European starlings, 110
evening grosbeaks, 18,
    51
extra pair phenomenon/copu-
    lation, 30, 113–15

feathers
  arrangement of, 66, 67
  collection of by male, 67,
    70–71, 165–67, 201–2
  experiment with, 65–66,
    67–74, 77, 96–98,
    100–101
  nest parasitization and,
    40
  play with, 65–66, 100–101,
    157–62, 203–4
  possible reasons for, 63–64,
    213–14
  provided by author, 28–29,
    70–71, 125–26, 132–33,
    142, 143, 153–54,
    191–92, 203–4
  removal of, 66–71, 73–74,
    77, 136, 155, 166–67,
    206
  toilet paper as substitute
    for, 29, 64
Finch, Robert, 216
finches, 194
  goldfinches, 34, 51
  purple, 51

fledging, 2–4, 6–7, 12–18,
    22–23, 82–87, 103–4,
    174–85, 206–7
flycatchers, great crested,
    110
flying squirrels, 145

geese, 49
golden-crowned kinglet, 51
goldfinches, 34, 51
grackles, 6, 51
great crested flycatchers, 110
grosbeaks
    evening, 18, 51
    rose-breasted, 5, 6
grouse, ruffed, 77

hairy woodpecker, 76
hatching. *See* eggs, hatching of
hawks
    broad-winged, 4
    Cooper's hawk, 3, 4
hermit thrush, 106, 107, 190,
    196, 199
herons, 64, 90, 91, 213
house wrens, 110
human presence, acclimating
    birds to, 52–53, 56. *See
    also* feathers, provided
    by author
hummingbirds, ruby-throated,
    5, 64, 74, 96

incubation, 1, 32, 44, 71–73,
    100, 135
individuals, focus on, xiii–xiv

intruders, battling and chasing,
    9–10, 21, 27, 41, 54–55,
    91, 95, 110–11, 150

juncos, 51
juvenile coloring, 54

kinglet, golden-crowned, 51

laying. *See* eggs, laying of
loons, 18

mating and mating behavior
    in 2012, 21
    in 2013, 30–32, 33, 34,
        36, 41
    in 2014, 61
    in 2015, 98, 99–100
    in 2016, 118–21, 123–25,
        127–28, 129, 132, 134
    in 2017, 151–55, 159
    extra-pair mating, 113–15
    mock mating, 94, 95, 129
McVey, Margaret, 90
merlins, 112, 165, 176, 197–98
migration, 105
mourning cloak butterflies, 106
mourning doves, 5, 51, 106,
    194

nest parasitization, 36–37,
    39–40, 121–22, 131,
    133–35, 172
nesting sites
    competition for, 25–26, 61,
        110

nesting sites (*cont.*)
  selection of, 51–54,
    188–89
nests
  guarding, 38–39, 41–44,
    78, 98, 109
  hygiene and, 22, 79
  materials for, xi–xiii,
    63–64. *see also* feathers
  preparation of, 28, 55–61,
    92–93, 117–19, 123,
    125, 152–56, 189–90,
    191–94, 195
Nicrophorus burying beetle,
    86–88
Nicrophorus pustulatus,
    86–88
northern flicker, 51
northern orioles, 6
nuthatches, 213
  red-breasted, 30

orioles, northern, 6
ovenbirds, 18, 116, 132, 190,
    197
owls
  barred, 51, 54, 116
  snowy, 89

Peterson, Roger Tory, 217
phoebes
  behavior of, 76–77,
    133
  hatching of eggs of,
    154
  nest of, 51

nest parasitization and,
    121–22, 129, 131–32,
    134, 136, 140
  return of, 106, 148
  song of, 61, 74, 107
punctuality, 187
purple finches, 51

raccoon, 87
ravens, 18, 209
red squirrels, 145, 209
red-breasted nuthatches,
    30
red-eyed vireos, 77, 196
red-winged blackbirds, 49,
    51
robins
  hatching of eggs of,
    154
  observation of, 5, 34
  song of, 51, 106, 107, 131,
    148
rose-breasted grosbeak, 5,
    6
ruby-throated hummingbird,
    5, 64, 74, 96
ruffed grouse, 77

sapsuckers, yellow-bellied. *See*
    yellow-bellied sapsuck-
    ers
scarlet tanager, 77, 196
sexton beetles, 86–88
sky space, defense of, 42
snowy owls, 89
song sparrows, 5–6, 51

sparrows
  English, 110
  song, 5–6, 51
  white-throated, 51
Spitzer, Paul R., 216
squirrels, 145, 209
starlings, European, 110
starvation
  brood reduction and, 90
  hatching times and, 39
  nest parasitization and, 115
  of nuthatch nestlings,
    30–31
  of swallow nestlings, 7–8,
    14, 19, 46–47, 182–83
swallows
  bank, 10
  barn, xii–xiii, 9–10, 90, 187,
    213–14
  cliff, xii–xiii, 10, 64, 187,
    213–14
  peaceful nature of, 9–10
  See also eggs; nests; tree
    swallows

tanager, scarlet, 77, 196
Taylor, Liam, 215–16
"third party" birds, 79–81,
    82, 114–16, 173–74
Thoreau, Henry David, xiii,
    47
thrush, hermit, 106, 107, 190,
    196, 199
toilet paper, as substitute for
    white feathers, 29, 64
torpor, 180

tree swallows
  battling and chasing in-
    truders, 9–10, 21, 27, 41,
    54–55, 91, 95, 110–11,
    150
  marking of, 134, 136
  return of, 49–50, 51–52,
    106–8, 148–50
  scientific studies of, xi–xii,
    209
  starvation of nestlings
    of, 7–8, 14, 19, 46–47,
    182–83
  See also eggs; nests
tufted titmice, 110
turkey vultures, 19
turkeys, 116, 148

Vance, Mark H., 217
vireos
  blue-headed, 51, 54, 90,
    196
  red-eyed, 77, 196
vultures, turkey, 19

warblers
  chestnut-sided, 18, 34
  yellow-throated, 18, 74,
    131, 196, 197, 199
waxwings, cedar, 5
Webb Lake, 49–50
Wheelwright, Nathaniel,
    215–16
white-throated sparrow, 51
winter, swallows during, 105,
    215–18

winter wrens, 51, 190, 191, 196

woodcocks, 51, 106

woodpeckers
    hairy, 76
    holes from, 114, 145, 213
    northern flicker, 51
    yellow-bellied sapsuckers, 51, 54, 61, 96, 106, 107, 132, 195

wrens
    house, 110
    winter, 51, 190, 191, 196

yellow-bellied sapsuckers, 51, 54, 61, 96, 106, 107, 132, 195

yellow-throated warblers, 18, 74, 131, 196, 197, 199